T0246032

Rumbles

Rumbles

A CURIOUS HISTORY OF THE GUT

The Secret Story of the Body's Most Fascinating Organ

Elsa Richardson

PEGASUS BOOKS

NEW YORK LONDON

RUMBLES

Pegasus Books, Ltd.
148 West 37th Street, 13th Floor
New York, NY 10018

Copyright © 2024 by Elsa Richardson

First Pegasus Books cloth edition October 2024

ISBN: 978-1-63936-724-5

10 9 8 7 6 5 4 3 2 1

Printed in the United States of America
Distributed by Simon & Schuster
www.pegasusbooks.com

For Frank

Contents

Introduction

Writing in 1794, Erasmus Darwin – physician, abolitionist and eventually grandfather to the founder of evolutionary theory – recounted the case history of a rather unfortunate patient. A girl of sixteen who, along with all the other awkward afflictions that puberty brings, was plagued by a remarkably vocal gut. Her bowels, Darwin recorded, 'almost incessantly made a gurgling noise so loud as to be heard at a considerable distance, and to attract the notice of those who were near her'. The noise, which 'never ceased a minute', caused understandable distress to the sensitive patient, drawing public attention to the body's innermost workings in a way that must have been especially excruciating for a teenage girl.[1] Most of us, even if we have not had to seek medical treatment, will have heard the same resonant tumult from within. Of the seventy-eight organs of the body, the stomach is certainly among the noisiest. While the kidneys filter waste without a peep, the lungs quietly puff away and even the steady beat of the heart can only really be heard by pressing an ear up close to the chest, the belly is notoriously outspoken. It growls, gurgles and grumbles. Its clamorous sounds are often a source of embarrassment, drain-like modulations that bubble up from the deep without warning, bringing with them news of digestion and defecation. Its rumbles disturb the hushed silence of libraries and echo through the civilised quiet of museums, interrupt important interviews and make unwelcome interventions into promising first dates. There is something almost obnoxious in the way the organ barges its way into our social world, drawing attention to the fleshy fact of the body and its most shame-inducing functions. Even the medical term for gastric noises – *borborygmus* – has an unappealing bilious ring to it.

To quieten his patient's obstreperous belly Darwin devised a specially tailored course of treatment: she was to swallow 'ten corns of black pepper' after dinner, take a daily dose of crude mercury and allow a 'small pipe' to be occasionally inserted into her rectum to 'facilitate the escape of air'.[2] This dispiriting prescription would seem to imply that they were dealing with a purely physical problem, but in his notes Darwin pointed to another possibility: an excessively noisy gut, especially in a young woman, was often a symptom of 'fear'. Long fascinated by the complex relationship between mind and body, Darwin wagered that his patient's unsettled mental state might produce gastric distress, trapping her in a dreadful cycle in which negative feelings upset the stomach, whose vocal protests then provoked even greater anxiety. According to this diagnosis, the gut was a uniquely sensitive organ capable of registering the slightest modulations of mood, but it also seems to have possessed a kind of social agency. In the case history recorded by Darwin, the talkative bowel was a problem not just because it caused physical discomfort, but also because its loud protestations exposed the murky inner workings of the viscera to the outside world.

Eating involves crossing the same border in the opposite direction, taking something from the outside into the deep interior of the body. This simple act has significant consequences because, as the historian Michael Schoenfeldt has observed, the incessant demands of appetite 'require the individual to confront on a daily basis the thin yet necessarily permeable line separating self and other'.[3] The everyday nature of tasting, chewing and swallowing obscures how momentous these small acts are not only in terms of sustaining life but also, as will become apparent, in shaping our identity. Demanding the attention of family doctors like Darwin, as well as dieticians, surgeons, gastroenterologists, public health officials and social reformers, the digestive system has long been central to how we understand the body, but the nature of its connection to our emotional, social, cultural and political lives is less clear.

Rumbles: A Curious History of the Gut asks how this noisy organ came to shape how we see ourselves and the world around us. Unearthing a fascinating gastric discourse from Ancient Greece to Victorian England, eighteenth-century France to twentieth-century North America, it draws together religious tracts, medical treatises, literary texts, anthropological studies and etiquette guides to argue that there is far more to the belly than might first appear. It recounts the complex history of the gut through a wildly diverse cast of characters: scientists – pioneering and wrongheaded in equal measure – Edwardian bodybuilders, hunger-striking suffragettes, philosophers, sandal-wearing vegetarians, Romantic poets, animal rights activists, psychoanalysts, witches, demons, royals, sanitary reformers, anatomists, microbiologists, novelists, city planners, playwrights and medieval alchemists. It is interested in all the different ways that we have conceptualised, visualised, animated and storied the gut. What metaphors have been called upon, what specialist lexicons have developed, what narrative strategies have we developed to understand the work of digestion?

Medicine is, perhaps, the most obvious place we might look for answers to these questions. This book visits laboratories, specialist clinics, anatomy theatres and lecture halls to spend time with a host of doctors beguiled by the gut's mysteries. The history of medicine has fascinating stories to tell, but the language of the gut can just as easily be found in plays, poetry and prose. Buried in the viscera, far from curious eyes and prying fingers, throughout history medical descriptions of its operations have relied as often on metaphor and simile as they have on precise clinical terminology. Even today, when advanced technologies have exposed the workings of the organ to detailed scientific analysis, the question of how best to communicate with the gut remains subject to contestation. It may reveal important information to the careful ear of a medical practitioner, specially trained to decipher meaning from its splutters and gurgles, but it speaks to all of us regardless of our professional qualifications. Digestion is a pleasurable, disquieting, messy and universal experience, over which we are all able to claim some measure of linguistic authority. While the workings of other organs – the heart's intricate assemblage of valves, atriums and ventricles or the networked neurons and firing synapses of the brain – demand clinical expertise, the digestive system has been formed by a lively public discourse in which specialist research jostles with a vast multiplicity of other ways of knowing, describing and inhabiting the body.

The stomach is, after all, an organ with which we are in near constant conversation. It demands to be filled several times a day and in doing so makes itself known to us in a way that is quite distinct from most of the rest of the body: we are conscious of its movements, while the work of other vital parts like the spleen, pancreas and liver continues well below the surface of our awareness. Though our sense of interoception, the brain's perception of the body, is formed by receptors attached to all the internal organs, it is the growling stomach that seems to speak the loudest. Perhaps its uniquely active presence goes some way to explaining why the gut has so often been personified, endowed with individual characteristics and the ability to express itself in language. In Geoffrey Chaucer's 'The Summoner's Tale', one of the twenty-four stories published as *The Canterbury Tales* between 1387 and 1400, the belly of greedy Friar John appears to speak for him, while *The Life of Gargantua and Pantagruel* (1564), François Rabelais' riotous Renaissance satire, features an 'Italian woman of low birth' whose stomach, when questioned, replies in a voice that is 'weak and tiny, yet always well-articulated, distinct and intelligible'.[4] Among the most erudite guts in this literary tradition is the narrator of Sydney Whiting's *Memoirs of a Stomach* (1853), a much-beleaguered internal organ who recounts his alimentary troubles and digestive trials for the reader. Plagued by the debilitating effects of nervous excitement and continental cuisine during his owner's college years, matters take a turn for the worse when he secures employment in the city. There, freed from familial and financial restraint, his master embarks upon a programme of regrettable overindulgence, as Mr Stomach complains: 'My belief is, if I had never passed the Temple Bar, I should at this moment have been at least ten years younger in feelings; but as it is, my constitution is undermined, and not with blowing up, but blowing out.'[5] The conceit of a gut that not only talks but talks back to expose the folly and greed of its owner so tickled Victorian readers that *Memoirs of a Stomach* became a bestseller translated into several languages. The popularity and persistence of this literary conceit suggests that digestion might play more of a role in shaping how we think about ourselves and the world around us than is usually acknowledged.

Even when it is not being creatively animated, the stomach has a lot to say because the organ and its processes are deeply embedded in the rhythm, culture and language of everyday life. Think of idioms like the familiar advice to follow a 'gut feeling' or the admission that you find something 'hard to stomach'. A well-known German saying has it that *Liebe geht durch den Magen* or love goes through the stomach, while in Italy *stare sullo stomaco* means that someone or something is annoying, they 'get on your stomach', and if a Swede advises you *is i magen*, to have ice in your stomach, you need to play it cool. Outside of well-known sayings, we might consider the ubiquity of digestion as a metaphor for learning, or the rhetorical link between aesthetic taste and gustatory pleasure, or the prominence of scatological images in the rich lexicon of obscenity as further evidence of the peculiarly lively life that the gut seems to enjoy outside of the body. In *The Second Brain* (1998) the biologist Michael Gershon describes the digestive tract as an 'open tube that begins at the mouth and ends at the anus [. . .] a tunnel that permits the exterior to run right through us'.[6] This image of the gut as a kind of opening, a crack where the world seeps in or a draughty corridor that it whistles down, is used by Gershon to make the case for a more nuanced understanding of the important ways that bodies and environments are entangled with each other. It is also a powerful illustration of how the gut mediates between the vast complexity of the exterior world and the sticky universe of interior life. Against the common conflation of mind with identity, this book argues that it is possible to read digestive processes – consumption, absorption, defecation – as other ways of making the self and making meaning in the world. As socially awkward as a loud belly might be, the rumblings of this garrulous organ can tell us a great deal about our shared culture and its history.

What Is the Gut?

Eating, seemingly the most mundane of acts, is in fact a rather complicated business. The digestive process requires a confederacy of different organs to work together harmoniously, entails the careful production of special chemicals and enzymes, and involves the remarkable assimilation of material from the outside world into the substance of the body. Digestion begins in the mouth with the release of saliva that – with the help of the teeth and the tongue – begins to break down food ready to be passed to the oesophagus, which pushes your meal downwards towards the stomach. The inner walls of this J-shaped organ are lined with glands that release gastric juice, while its muscly outer layers contract to mix this liquid with the food consumed to make what is known as chyme. This substance then moves into the first part of the small intestine, the duodenum, where bile from the gallbladder and enzymes from the pancreas break the material down further. The small intestine also features millions of tiny villi – finger-like structures with blood vessels inside – that channel the nutrients released by digestion into the bloodstream. After most of your meal's goodness has been absorbed and transported around the body, what remains is moved on to the large intestine which uses bacteria to complete the final stages of digestion. Over several hours yet another transformation takes place as water is absorbed from the chyme and it becomes semi-solid faeces, ready to be transferred to the rectum and finally expelled via the anus. This multi-step procedure is coordinated by the brain, nerves and several dedicated hormones, which send signals around the body preparing each part to be ready to play its assigned role at exactly the right moment.

Over the last few decades there has been an explosion of research into the workings of the gut: many scientists now believe it plays a far larger role in human physiology than was previously realised, influencing not only gastrointestinal health itself, but also obesity, metabolic health, the immune system, cognitive function, mental illness and neurodegenerative diseases. Much energy has been devoted to exploring two of the organ's most fascinating characteristics: namely, its peculiarly intimate relationship to the brain and the wild alliance of bacteria, archaea, viruses and fungi that it houses. Scientists have found that the gut communicates with the brain through several key routes in the body: the immune system, the vagus nerve that controls heart rate and digestion, tryptophan metabolism, which is associated with ageing and inflammation, and the enteric nervous system that governs gastrointestinal behaviour.[7] Elsewhere, explorations continue into one of the most diverse habitats on the earth: the human microbiome. Our bodies play host to trillions of microbes. These are spread across locations as diverse as the eyes, lungs, skin and mouth, but by far the largest community is to be found in the stomach, colon, small and large intestines. Brought to light by advances in genetic sequencing that revealed the extent of its heterogeneity for the first time, the gut microbiome begins to develop in the womb and appears to have a hand in everything from gastric health and metabolic function, to sleep and mental wellbeing. In laboratories around the world researchers are attempting to understand the scope and significance of these connections. At the Max Planck Institute for Metabolism Research in Cologne scientists are using new imaging technologies to map the neuronal signalling pathways of metabolism; over at the University of Cork progress is being made in understanding the link between cognitive function and microbes, with the tantalising possibility that altering the flora of the gut may help reverse age-related deterioration of the brain; the Laboratory of the Gut Brain Axis at Mount Sinai Hospital in New York is exploring the possible role that the microbiome might play in treating patients with multiple sclerosis; and in Belgium at KU Leuven psychiatrists have teamed up with gastroenterologists and nutritionists to investigate the links between digestion and mood. Wide-ranging, varied and often involving cross-disciplinary collaboration, the riddle of the gut has produced one of today's liveliest fields of research.

Research into the microbiome and advances in our understanding of biochemical signalling between the gastrointestinal tract and the nervous system have also provoked popular interest in the relationship between the stomach and the mind. Books like Giulia Enders' bestselling *Gut: The Inside Story of Our Body's Most Under-rated Organ* (2015) and Alanna Collen's *10% Human: How Your Body's Microbes Hold the Key to Health and Happiness* (2015) have told largely biological stories about the gut–brain axis, while texts like Gerd Gigerenzer's *Gut Feelings: The Intelligence of the Unconscious* (2007), Emeran Mayer's *The Mind–Gut Connection* (2016) and Alana and Lisa

Macfarlane's *The Gut Stuff: An Empowering Guide to Your Gut and Its Microbes* (2021) urge readers to transform their lives by tapping into the wisdom of the stomach. These books have, in turn, spawned a booming wellness industry dedicated to selling better digestion in the form of probiotics, juice cleanses and fermented foods that promise to revolutionise physical and emotional health from the belly up. All this attention has encouraged us to see the gut in a fresh light: as an organ of infinite complexity, containing multitudes of mysterious microbial life, deeply networked into the rest of the body, and strangely entangled with the intricacies of our psychic lives. These modern revelations are, though undeniably valuable, almost entirely the products of biomedicine, with the result that the gut we encounter across academic papers, self-help books, newspaper articles and promotional materials is one described and defined almost entirely by science. There are, however, other ways of imagining, representing and experiencing the body.

Take the word 'gut', for instance. A narrow definition of the term would have it refer only to the portion of the lower alimentary canal that connects the pylorus and the anus, but its common usage is far more expansive. It is often employed to denote the whole of the digestive system, a trend that has been encouraged by the popularisation of scientific research into the gut–brain axis that uses it as a more general descriptor. Look outside of medicine and its uses proliferate even further: *the gut* is the abdomen; *the guts* are the innards of the abdominal cavity, the bowels and entrails; *to gut* is to remove those entrails; *a gut* is a particularly fleshy belly, or even a person thought to indulge gluttonously. Away from the body *the guts* can refer to the interior of an object or to the key substance of an argument. We also use it to describe certain character traits and behaviours: to *have guts* is to be brave, to be *gutless* a coward; you can *hate a person's guts* and threaten to *have their guts for garters*; working hard might involve *sweating one's guts out* or *busting a gut*.[8] As these diverse applications suggest, the gut exists not only as a part of the anatomy, but also as a powerful metaphor capable of producing and communicating meaning in language. Fascinated by the belly's vigorous linguistic life, this book is committed to uncovering how ideas around consumption, digestion and defecation shape the world and our encounter with it. Where biomedicine views these processes as largely neutral functions of the body, *Rumbles* argues that they are in fact bound up with understandings of gender, race and class. What we eat, the foods we find palatable, those we find disgusting, the diets we follow, the practices we adopt around sanitation and the feelings we attribute to waste are all freighted with heavy social, cultural and political baggage. At stake here are big questions around subjectivity, technology, ideology, sexuality, spirituality, nationhood and identity, questions that lie far beyond the scope of science. To find answers it is necessary to turn instead to history.

Does the Gut Have a History?

Gut health seems like a peculiarly modern obsession.

Alongside natural yogurt, fermented foods and dedicated cookbooks like Jeannette Hyde's *The Gut Makeover* (2015) and Eve Kalinik's *Be Good to Your Gut* (2017), it is now possible to undertake a faecal microbiota transplant – colloquially known as a 'poo transplant' or 'trans-poo-sian' – a procedure that promises to restore gut flora, reinvigorate the digestive system, help fight infection, and cure all manner of bodily ills. The growing demand for alternative therapies and the absorption of some of these into mainstream medicine – in 2022 Australia became the first country to grant regulatory approval to faecal transplants and others around the world show sign of following suit – speaks to the remarkably prominent position enjoyed by the stomach today. The organ's ascent can be attributed, in part, to the work of individual scientists, the advent of new technologies and the emergence of dedicated academic fields like neurogastroenterology. Ground-breaking as these developments are, they do not fully account for the immense purchase the gut has on the modern imagination. There is something, as this book will argue, cultural at stake here.

Our current preoccupation with our bellies may stem, in part, from the fact that they seem to be giving us more trouble than they used to. Over the last few decades there has been a marked increase in the number of people recorded as suffering from chronic digestive problems. Inflammatory conditions like Crohn's disease and ulcerative colitis are on the rise: research carried out by the Bill & Melinda Gates Foundation found that in 1990 there were 3.7 million cases reported globally, a figure that rose to 6.8 million in 2017.[9] A recent study by the European Parliament painted a fairly grim picture: digestive cancers represent nearly 30 per cent of all cancer-related deaths, irritable bowel syndrome (IBS) afflicts a substantial portion of the population and yet remains stubbornly difficult to diagnose, and gastrointestinal disorders threaten to overburden healthcare systems in the future.[10] Asked to explain these figures, experts have pointed to the possible influence of various factors from the overprescription of antibiotics and the use of personal sanitary products that kill all bacteria – good and bad – on the skin, to environmental toxins and the overconsumption of processed foods. The debilitating symptoms of conditions like IBS are, according to some, one result of the developed world's attachment to extreme hygiene and detachment from the so-called 'natural' world. We have, it appears, discovered too late that the belly is a microbial garden to be carefully tended, not aggressively cultivated with pesticides and chemical fertilisers. The gut's gurglings appear to speak directly to the concerns of the present: to the increasing incidence of chronic illness and worries about processed foods, to anxiety, depression, and the hope that the solution to highly complex problems might lie in simply eating better.

Yet by imagining the gut as a palimpsest onto which the concerns and anxieties of the present day are traced, we are following the example set by past physicians, politicians, writers and philosophers who were similarly confident that the organ revealed something unique about their time. From Puritan minister John Paget's condemnation of the late sixteenth century as an era that put the 'belly' before spirit, to the seventeenth-century doctors who identified bad digestion as the price to be paid for their thriving civilisation, to the Victorian moralists worried that the stomach was ill-equipped for city living and the early twentieth-century psychoanalysts who interpreted alimentary anxieties as symptoms of the sexual neurosis engulfing modern society, the digestive system has often been called upon to encapsulate the tone and tenor of a particular moment in time.

Which is, in part, why it might be possible to conceive of the gut as something with a history. But if so, what kind of history might that be? Certainly, it does not align with the definition of 'history' offered by the *Oxford English Dictionary* as a 'narrative constituting a continuous chronological record of important or public events' and it cannot be said to have a history in the same way that major events like, say, the Reformation or the Second World War do.[11] It could be argued, quite reasonably, that the organ is itself unchanging, that it is a fact of human anatomy whose consistency renders it timeless. Yet this would be to accept that not only does it have no meaning beyond the strictly biological, but also that the 'biology' of the gut has been understood in the same way across time and space. In truth, knowledge of what exactly the organ is and how it functions has always been subject to much contestation. Throughout the Middle Ages, for instance, the Latin *venter* was used to describe both the gut and the womb, so that the belly was both digestive and procreative; well into the nineteenth century physicians debated whether digestion was a mechanical or chemical process; and as we have already seen, what body parts are thought to constitute 'the gut' changes from context to context. Along with these shifting definitions, the images we have used to describe the gut have also changed. Today we speak in ecological terms – adverts for probiotics encourage us to nurture the microbial garden within – but in early modern Europe the stomach was imagined as being more alike to the bustling kitchen of a great country house, while eighteenth-century physicians fussed over it as a nervously afflicted invalid and through the Victorian period it was frequently condemned as an irascible foe, an enemy within implacably opposed to its owner's comfort.

Bodies are more than just their biology, they are made by language, culture, environment, and a myriad of other possible components; they exist in a context and are defined by the fine peculiarities of that context. Moreover, disciplines like biology are also themselves historical, methods of enquiry that change across time and produce different versions of the human body along the way. In his book *The Politics of Immunity* (2022) Mark Neocleous proposes that not only does the body have a history, but also that it moves through distinct historical epochs. From the mechanical body conjured by new technologies in the seventeenth century, to the mimetic body provoked by the eighteenth century's fascination with clockwork and automata, to the digital body of the twentieth century's information age and the genetic body created by the twenty-first, scientific knowledge of the body is always formulated with other kinds of discourse in mind.[12]

Working from a similar premise, this book proposes that we can think about the history of the gut in two separate but related ways. Firstly, it argues that the organ itself has a fascinating history – involving medical breakthroughs, technological innovations, philosophical experiments, literary imaginaries and religious fervours – waiting to be uncovered. Secondly, it contends that though the story of the gut may not be that of the Second World War, it does reveal that what we tend to think of as the stuff of history – international conflicts, political revolutions, scientific discoveries – has often been bound up with the needs of the body and the demands of the belly.

In recounting the curious story of the gut, *Rumbles* draws from distinct but related fields of academic research: cultural history and the history of the emotions. It is possible to read this book as a history of medicine, but its focus lies not in recounting the lives of great physicians or rehashing tales of well-known breakthroughs, but rather with where medical ideas about digestion rub up against lived experience, spiritual values, art, poetry, folk beliefs, social ritual and customary knowledge. These areas of investigation are territories usually associated with cultural history, a way of looking at the past that attends closely to the shape and cadence of everyday life. As Raymond Williams, one of the pioneers of this approach to history, put it: 'Culture is ordinary [. . .] every society has its own shape, its own purposes, its own meanings. Every human society expresses these, in institutions, and in arts and learning.'[13] On the one hand, cultural history takes the 'ordinary' seriously: it examines what people ate, how they had sex, what they did for fun, what books they read, how they felt about family life and so on, with the aim of coming to some understanding of what it was to live a life in a time other than our own. On the other hand, however, cultural history also resists the truism that people in the past were 'just like us' by making strange the commonplace, the natural and the taken for granted. When, for instance, we learn that the concept of child-hood has changed dramatically over the last four hundred years, from the pre-industrial world where the transition to adulthood began at age six to the modern day where the right to be considered a child until sixteen is enshrined in law, then something that might have seemed immutable suddenly reveals itself to be contingent. *Rumbles* harnesses this same revelatory power to make us look again at the seemingly unremarkable, unchanging work of the diges-tive system and to perhaps spot something unexpected or perplexing there.

The history of the gut is also an emotional one. Much of our current enthusiasm for the organ has been generated by emerging insights from the world of science into the role it might play in shaping our emotional lives. Recent studies into neuroendocrine signalling have identified the micro-biome as a key site of the production of so-called 'happy hormones' like dopamine and serotonin, while other experiments have explored the psychosomatic origins of common gastric conditions like IBS, whose sufferers often report depression, stress or anxiety alongside abdominal pain, bloating and discomfort.[14] This research, though innovative in other ways, relies on an understanding of emotions as straightforward biological reflexes, consistent across time and space. The history of emotions, a discipline in ascendancy over the last two decades, offers a different way of thinking about the relationship between mind and belly.[15] Interpreting emotions as historical constructs rather than universal constants, scholars working in this field have explored how our embodied experiences are shaped by social pressures, cultural influences, intellectual fashions, religious practices and political whims.

Taking a similar approach, this book complicates the dominant biomedical version of the gut–brain connection, suggesting instead that because this relationship has been largely conceptualised outside of the realm of scientific research – by literature, art, etiquette books, self-help guides and even demonology – it might be better approached using the interpretative tools of the historian. Mapping the curious history of the gut, *Rumbles* seeks to influence scientific research by drawing attention to the importance of terminology, imagery and metaphor in shaping our conception of the body, and by pressing those committed to exposing the link between digestion and the emotions to consider how the historical nature of terms like 'happiness', 'sadness' and 'anxiety' might complicate any simple understanding of that connection. This is with the aim, in part, of fostering a more varied, more expressive, more emotional language of the gut, that might better encompass our shifting relationship to diet, digestion and waste. Throughout history the gut has exercised more of an influence on how we have structured knowledge of ourselves and the world around us than we previously thought.

Emotions, Work, Time and Politics

The story *Rumbles* wants to tell is, in one sense, a broad one that traverses a lengthy period from the ancient world to the present day, calling upon a polyphony of different images, texts and ideas to unearth the complex history of the gut. It is also, however, fairly conservative in geographical terms, rarely straying far from Britain, continental Europe and North America. This narrow focus is partly a product of authorial expertise, or rather lack of expertise, regarding the complex histories of the rest of the world, but it is also because there is something quite specific about the way eating, digesting and defecating came to be entangled with identity in the Western world. At the heart of this book is a fascination with the way digestion, as a bodily process, metaphor and cultural discourse, has helped to make the modern self. For instance, it explores the way that conceptions of nationhood and subjecthood have been formed through the work of the gut. Citizenship has long been connected to consumption and digestion in unexpected ways: consider, for example, the role of certain foods in shaping national identity – steak and chips in France or roast beef in England – or the way that perceived outsiders are demarcated as such with regard to their culinary methods. Going further, we might also consider the self-regulating practices that conflate dietary control with national fitness or the way ideas about purity traffic between the alimentary and the moral, or how taboos around bodily waste have so often been used to divide the 'civilised' world from the 'primitive'. Cultural practices and shared narratives around the gut have long been used to bolster the West's sense of itself as exceptional and superior.

The gut also troubles individual and collective vanities. Case in point is Darwin's unfortunate young patient, whose burbling bowel brought her into daily conflict with the societal expectations of the period. She may have been as meek and politely spoken as befitted a girl of her class, but her stomach was a boorish loudmouth whose unruliness threatened to undermine its owner's careful performance of proper 'femininity'. Many of the tales collected in this book feature guts that refuse to behave. From the third-century philosopher who warned that the stomach was uniquely vulnerable to demonic possession and the medieval doctors who fretted over the poisonous vapours fermenting in the depths of the bowel, to the Romantic poets plagued by the bowel's melancholy 'blue devils' and the nineteenth-century office workers brought low by the perils of indigestion, the gut has been consistently depicted as unpredictable and ungovernable. With this in mind it is perhaps not surprising that our relationship to the organ has so often been characterised as a struggle for control. The alimentary tract may be, as Gershon has it, a 'tunnel that permits the exterior to run right through us', but it is one with closely guarded entrances and exits. The management of these portals, what is permitted into the body and what must be expelled from it, is an activity fraught with social pressure. Consider the opprobrium levelled at 'unhealthy' eating habits or the proliferation of taboos around the toilet: digestion is a process that comes heavily weighted with anxieties around self-discipline, individual agency, and the degree of control it is possible to exercise over our bodies. This book is interested, then, not only in how digestion has helped to shape the self, but also in all the ways that it has threatened to unmake it.

Rumbles begins its curious journey through history with the mind. Modern science has, as we have seen, devoted itself to uncovering the mysteries of the gut–brain axis, but the idea that the belly might have some part in the life of the mind is far from new. Medical thinkers from Ancient Greece onwards have insisted that the stomach is the true source of our emotions, others have held the digestive system responsible for bad mood and some have blamed our passionate tumults for pangs of gastric distress. Over centuries of philosophical tracts, religious treatises, scientific research, dietary advice and poetical musings, the organ becomes implicated in conversations that seem, at first glance, far removed from its workings. Across several chapters, this book explores the role played by the gut in debates concerning the relationship between the immortal soul and the mortal flesh, what it means to be 'civilised', where intelligence resides in the body and the nature of consciousness itself.

Thinking about the mind, then, has often involved investigating the stomach, but it is not the easiest organ to examine. Not only is it buried far down in the viscera away from impertinent fingers and prying eyes, it also only really makes sense when it is in motion. Unlike other organs that hold their shape when they are removed from a dissected cadaver like the heart and the liver, the stomach quickly comes to resemble a deflated carrier bag. It is defined by the toil of digestion, by the steady labour of peristaltic waves moving food from mouth to intestine. This goes some way to explaining why it has so often been characterised as the 'worker' of the body or had its operations and functions understood in relation to the world of work. Moving on from the emotions, *Rumbles* spends several chapters with the history of the working gut. Starting with an enquiry into how the literal 'work' of the stomach has been imagined and the technologies that have been developed to witness it in action, the book then considers the sometimes fraught relationship between the work of the brain and the work of the stomach, before examining how concerns over the possible link between white-collar labour and gastric distress came to define the modern workplace.

The gut is entangled with the history of work and nowhere is this more obvious than in the question of lunch. In addition to telling the story of the midday meal, beginning with its origins in nineteenth-century Britain, *Rumbles* also argues that the institution of lunch*time* is itself significant. The institution of a set hour of the day to eat speaks to the broader standardisation of daily time, a process that gathered pace during the Industrial Revolution where it extended beyond the production line to the body of the worker. Regulated breaks – hard won by union activism – established a particular rhythm for the digestive system but, as the following chapters explore, the stomach has a relationship to time that extends beyond the Victorian factory. From ancient Babylonia, where it was possible to glimpse the future in the entrails of sacrificed virgins, to the evolutionary biologists who condemned the stomach as a primitive vestige lurking in the modern body, the gut has often been framed in temporal terms.

Examining its past, present and future, it quickly becomes clear that the time of the gut is no neutral matter. Indeed, how we think about the tempo of the digestive process and its products is often bound up in the broader politics of gender, race and class. Taking up the gut as a political organ, the closing chapters of this book examine where the inner workings of the body rub up against the institutions of the state: from the role of the stomach in shaping the body politic to the actions of hunger-striking suffragettes who harnessed the gut to demand the right to vote. Beyond the empty stomachs that have driven revolutions and shaped world-changing events, *Rumbles* is also fascinated by – to use a term coined by the American food historian Frederick Kaufman – 'gastropolitics', namely the way that the language of appetite and digestion has come to shape our thinking around what is virtuous, moral and valuable, and what is not.[16] This is especially visible in the history of dieting and in the loaded language that gathers around the 'issue' of weight. Digging into the persistent characterisation of the 'fat' belly as a disruptive, unruly organ, the closing chapters of this book reflect on the gut's connection to acts of political disobedience and bodily rebellions.

Food plays many roles in our cultural and social worlds. What we eat helps to establish individual and collective identities, marking out subtle distinctions in class, gender and race, and helping to set the boundaries between animal and human, nature and civilisation. There is a great deal that could be said about the various complexities of this relationship, but *Rumbles* is especially interested in the question of restraint, with all the ways that we have tried to exercise control over one of the most pleasurable bodily experiences. Diet recurs as a theme throughout this book, but its history turns out to be a lot weirder than expected. We meet a seventeenth-century hermit convinced that the only true path to God is to eat only herbs and roots; physicians assured of the link between poor diet and bad dreams; Victorian vegetarians convinced of the danger posed to belly and mind by meat; and a Russian-born microbiologist who believed the secret to a long life lay in a daily bowl of yogurt. Exemplifying what the historian Corinna Treitel describes as the way that dietary knowledge has typically been formed in 'messy and non-hierarchal ways' through processes characterised by the 'dual dynamics of critique and co-optation', *Rumbles* is interested in the gut as a site of persistent lay experimentation.[17] Choosing what to eat is an everyday intervention into health and, prior to antiseptics, antibiotics and anaesthetics, diet was by far the most sensible approach to curing bodily ills. Moving the gut from the periphery to the centre of medical history, this book argues that the processes of consumption and digestion are essential to understanding embodied experience.

Writing along similar lines, the philosopher Stanley Cavell observed that the 'body has (along with the senses) two openings, or two sites for openings, ones that are connected, made for each other' and that no one has yet formulated 'what the right expression is of our knowledge that we are strung out on both sides of a belly'.[18] Charting the curious cultural history of the gut, taking in medical triumphs and failed experiments, moving through Renaissance dinner tables and workplace canteens, cramped studies and hi-tech laboratories, reading poetry, letters and novels along the way, this book is a wide-ranging exploration of what it means to be a digesting animal. Fascinated by all the ways that we have experienced and processed the world, *Rumbles* examines how modern selfhood has been shaped by the gut.

MIND

George Cheyne had become, by his own admission, rather out of shape. He had never been a particularly fit man, but by the early 1720s his health had entered a state of rapid decline. Having moved to London to build a medical practice, the young doctor had found himself 'constantly dining and supping in Taverns' with important new contacts, and this period of chronic overindulgence eventually took its toll.[19] Finding himself lethargic, melancholy, covered in 'scorbutic ulcers' and unable to walk unaided, Cheyne put himself on a diet. The new regimen featured only '*Bread*', '*Seeds* of all Kinds' and 'tender *Roots* (as *Potatoes*, *Turnips*, *Carrots*)', with fresh milk and water to replace wine. Finding that this 'vegetable diet' improved not only his bodily health but also his general mood, he began to recommend it to patients as a solution to low spirits as well as a cure for common ailments like indigestion and gout.[20] Cheyne took up the question of what to eat and what not to eat most famously in *The English Malady* (1733), an influential treatise that attributed the nation's melancholic temperament to bad weather, polluted cities and gastronomic excess. The problem, according to Cheyne, was that British society was slowly being deformed by its own prosperity and imperial successes. Having 'ransack'd all the Parts of the *Globe*', the country now risked being brought down by luxury, greed and gluttony.[21] Indulging in wine, cheese and red meat was not only injurious to the body, he argued, it was also a sure route to nervousness and low spirits.

Writing at the beginning of the eighteenth century, as the empire expanded, overseas trade intensified and cities boomed, Cheyne fretted at the fate of the nation's bellies in a time of such great excitement and upheaval. Insisting on a close connection between the operations of the stomach and the workings of the mind, *The English Malady* urged readers to attend more carefully to the former to better support the latter. Though this idea – that how we feel is linked to what we eat – seems familiar, here it remains tangled up with the peculiarities of a moment in time that sits at significant remove from our own. While scientific accounts of the gut–brain axis tend to envision this attachment as rooted in biology and consistent across time, the opening chapters of this book are devoted to unearthing its rich and strange history. The story that emerges involves philosophers, psychologists, poets, dieticians, cartoonists, witches, royals, a hermit and one small brown dog, and it reveals that many of the difficult questions we have asked about what it means to be human – what intelligence is, where emotions come from, how the brain relates to the rest of the body, whether 'mind' is a uniquely human possession – come back again and again to the belly. Following Cheyne, who committed himself to transforming his emotions by changing his diet, it takes seriously the idea that the gastrointestinal tract might have some hand to play in shaping our complex inner lives.

1

The Intelligent Gut

We have not one, but two brains. This remarkable quirk of human anatomy was discovered in nineteenth-century Breslau by Leopold Auerbach, a German neuropathologist who developed a method of staining cells that made it possible to observe the microscopic structures of the nervous system for the first time. His major discovery was the myenteric plexus or Auerbach's plexus, a group of cells that direct the movements of the gastrointestinal tract, which include neurons like those found in the brain. These neurons form part of an army of 100 million lodged in the alimentary canal, capable of operating independently of the central nervous system and responsible for much more than just the daily grind of digestion. Today scientists point to the existence of cells that allow the gut to speak directly to the brain, and those working in the emerging field of neurogastroenterology have suggested that the stomach might have a role to play in cognition. This cutting-edge research speaks to the familiar advice to 'follow your gut' or 'trust your gut instinct', which rests on the assumption that the stomach might know something the brain does not. Or that, freed from the rationalising tendencies of the mind, emotions might be felt more acutely, more authentically, in the gut. This association is more than metaphoric – medical thinkers throughout history have identified the stomach as a peculiarly learned organ. Modern science is uncovering new facets of the gut–brain connection all the time but, as this chapter will explore, the idea of the thinking belly has long shaped debates about the nature of intelligence itself.

Much of our current fascination with the gut has been fuelled by a growing appreciation of the myriad ways that its workings are entangled with those of our higher cognitive functions. Neurologists have been tracing the path of the vagus nerve that runs from brain to bowel and elsewhere biologists mapping the microbiome have discovered that bacteria living in the intestines help to regulate the production of the biochemicals responsible for stimulating neuronal growth. Science has positioned itself at the forefront of this revolution in how we understand intelligence, but many of the questions it has raised – where is intelligence located in the body? What are its limits? What differentiates human from animal? – have been taken up with equal enthusiasm by philosophy, literature and popular culture. Moreover, while scientific research is often presented as objective and value-free, it is undertaken within and shaped by a particular social context. The work of these subtle forces can be hard to discern in the present, but by taking a historical view it is possible to conceive of the gut–brain connection as scientific knowledge that has been produced in relation to different kinds of discourse and different conceptions of the body. *Rumbles* begins its curious history by digging into some of the stories that lie behind the biomedical discoveries to uncover where the idea of an intelligent gut has bumped up against understandings of what it means to be human.

The modern history of the gut–brain connection begins with a small brown terrier dog. Around the turn of the twentieth century the Department of Physiology at University College London played host to a series of experiments that would not only lead to the discovery of hormones, but also reveal the true scope of the enteric nervous system. Though it was Auerbach who, peering at a section of bowel through his microscope in 1863, first spied the thick bundle of nerves nestled in between layers of muscle, it was only with the work of two British scientists – William Bayliss and Ernest Starling – that the significance of the myenteric plexus finally began to be understood. Close friends, collaborators and eventually even brothers-in-law, the pair conducted their investigations using anaesthetised dogs.[22] One of those unfortunate canines, a rather scruffy creature with alert eyes and soft ears, would become the subject of immense public controversy, a high-profile trial, riots and even a specially commissioned statue.

Initially interested in the nature of the relationship between the nervous system and pancreatic secretions, Bayliss and Starling discovered that, contrary to long-held orthodoxy, the former did not influence the latter. Instead, it appeared that the pancreas was encouraged to produce digestive juices by chemical messengers that originated in the walls of the intestinal lining, whose communications were delivered through the bloodstream. Drawing on this research in a lecture to the Royal Society of Physicians in

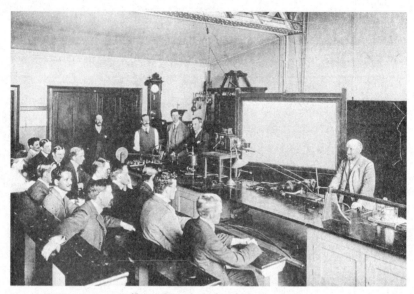

Figure 1: William Bayliss (right) demonstrating a
reconstruction of the 'Brown Dog' experiment, 1903

1905, Starling coined the word 'hormone' from the Ancient
Greek ὁρμῶν, 'to arouse or excite', to describe how 'activities and growth in different parts of the body' could be
stimulated by the excretions of seemingly remote organs.[23]
One of the most startling possibilities raised by this revelation was that the rule of the central nervous system
– the brain and spinal cord – might be subject to contestation by the remoter principalities of the body. To test this
hypothesis, the collaborators turned to their preferred test
subject, a small dog (Figure 1). Having anaesthetised and
sliced open the dog, the scientists first isolated and discon-
nected the nerves that linked the intestines with the brain.
They then proceeded to inject the animal with hydrochloric acid, which mimicked the effect of gastric movement.

Remarkably, even though the essential nervous connections had been severed, these movements still prompted the pancreas to begin secreting digestive juices. This result demonstrated that the bowel possessed, what Bayliss and Starling described as, its own 'local nervous mechanism'.[24] In other words, the gut could act independently of the brain.

Around the same time, over at Trinity College, Cambridge, the physiologist John Newport Langley was using nicotine to map out the autonomic nervous system (ANS), responsible for regulating heart rate, blood pressure, respiration, sexual arousal and digestion, when he discovered something unexpected.[25] Surveying the thousands of ganglia that line the stomach, small and large intestines, pancreas and gallbladder, he observed that these seemed to function unassisted. Writing in the 1900 edition of *Schäfer's Textbook of Physiology*, Langley was the first to describe the 'enteric nervous system' (ENS).[26] This intricate neuronal network was, he argued, not simply a subdivision of the ANS, rather the gastrointestinal tract appeared to be a wholly autonomous kingdom buried deep within the viscera.

The idea of the self-governing gut engaged scientists around the world. The American physician Byron Robinson published a whole book on the topic, *The Abdominal and Pelvic Brain* (1907), based on his extensive research into the abdominal nervous system, in which he described the gut as the 'automatic, vegetative, the subconscious brain of physical existence'. Where in the 'cranial brain resides the consciousness of right and wrong [. . .] in the abdomen there exists a brain of wonderful power maintaining eternal, restless vigilance' over the internal rhythms of the body.[27] Alongside Leopold Auerbach, the Spanish neuroanatomist Santiago Ramón y Cajal developed his own method of staining with silver that made the wild variety of intestinal neurones newly visible, while the Russian neuroscientist Alexander Dogiel explored the intricate structures of its neurotransmitters.

The efforts of this group of scientists, working independently of each other throughout the nineteenth century and into the early decades of the twentieth, laid the ground for the modern fascination with what Michael Gershon has described as the 'second brain'.[28] Gershon made a claim for the existence of this so-called second brain based on the self-governance of the gut, and the fact that it is a 'site of neural integration and processing' that 'can elect not to do the bidding of the brain or spinal cord'.[29] Unlike most other parts of the body, which would cease to function if their links to the central nervous system were severed, the peristaltic reflex – responsible for the wave-like muscle contractions that move food through the digestive tract – can operate independently. Recent research in the field of neurogastroenterology suggests that the ENS might be capable of even more than we currently credit it with: studies from Michigan State University have uncovered how enteric glia, non-neural cells thought only to support the activities of neurons, actually work independently to guide peristalsis. Others have pointed to the influence of the microbiome on mood and in 2020 a team of German scientists asked whether the gut might have the capacity to learn and memorise new information.[30] Biologists based at Flinders University in Australia have even argued that the gut is, in truth, the first brain, having evolved in humans long before the grey matter in our heads.[31]

Part of what might make the idea of a second brain so appealing, to scientists and lay people alike, is the way that it makes room for the possibility that the work of thinking might occur elsewhere in the body. That the belly might be the first to register a feeling – the stomach that flips when your crush walks into the room and sinks when they don't say hello – or that the gut might process information from the exterior world ahead of cognition – the quiver of the intestines that sometimes signals danger, for instance – reflects something of our embodied experience that has been, until recently, largely absent from mainstream medical accounts of the mind–body relationship. We are captivated by the notion of the thinking belly, in part because it seems to more accurately reflect how processes of attention, awareness and perception necessitate parts of the body other than the brain. Which raises the question: why has intelligence, at least in the Western world, for so long been viewed as the exclusive preserve of the cerebrum? How did the first brain come to dominate?

To find answers it is necessary to look back to the seventeenth century, to one of the foundational myths of modern medicine, an idea so powerful that it continues to shape not only how we imagine ourselves, but also how we conceptualise the inner lives of others, both human and non-human. In *Discourse on the Method of Rightly Conducting the Reason and Seeking Truth in the Sciences* (1637), the French philosopher and mathematician René Descartes formulated a theory of knowledge based on the application of radical doubt. Because the senses with which we perceive are so easily deceived it was impossible, he conjectured, to have any absolute certainty about the true nature of the world, beyond the fact of his own questioning mind. Hence his famous formulation: 'I think, therefore I am.' Asked to define what it is to be him, Descartes concluded only that he was 'a thing which thinks. What is a thing which thinks? It is a thing which doubts, understands, conceives, affirms, denies, wills, refuses, which also imagines and feels.'[32] Importantly, that 'thing which thinks' was considered to be wholly distinct from the body, as the latter – like everything else – could only be sensed because there was a mind there to sense it. From this initial observation, Descartes described the mind and body as existing in a dualistic relationship with one another, as entities, one immaterial and the other material, that interacted closely but remained fundamentally distinct. There was, for Descartes, certainly no 'second brain' lodged in the gut. The body was, he wrote, 'nothing but a statue or machine made of earth', a mechanical assemblage responsible for everything from respiration and digestion to sensory perception.[33]

Buried within this remarkable apparatus, setting it in motion and guiding its movements, was what Descartes described as the 'principal seat of the soul': a reddish gland, about a third of an inch long, squished between the two hemispheres of the brain (Figure 2).[34] In the 1620s, prompted by a fascination with the mysteries of the optic nerve, Descartes added medical science to a growing list of his scholarly pursuits that included geometry, physics and algebra as well as philosophy. He undertook a programme

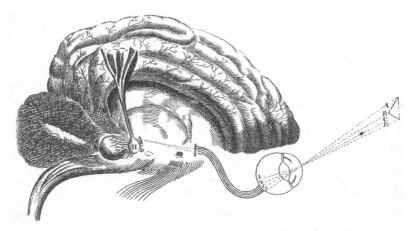

Figure 2: The pineal gland, as depicted in the 1662
Latin edition of Descartes's *Treatise on Man*

of study in the field of anatomy, focusing his attention on
the inner structures of the eye and attempting to understand their connection to processes of visual perception
in the brain. It is perhaps not surprising, given the hours
he spent rooting around in all that grey matter, that he
would take up the tools of the anatomist to uncover the
ineffable mysteries of the human mind. Writing in *Passions of the Soul* (1649), Descartes claimed that having
dissected countless specimens, he had 'discovered that
the part of the body in which the soul immediately exercises its functions is not in any way the heart; nor is it
either the whole brain, but only the most internal of its
parts, which is a certain very small gland'.[35] Also known
as the conarium, the pineal gland was, he insisted, significant because of the centrality of its location in the brain,
situated at a nexus of lively neural pathways, and its sin-
gularity. This was deemed to be important because of the

seeming contradiction between our doubled organs of sense – eyes, ears and so on – and our inability to perceive or think about more than one thing at once, which suggested that there must be some place where impressions were collated. As the pineal gland is the only part of the brain that is not double, Descartes reasoned, it must be the site at which images from the sensory organs can combine and thus represents the source of 'common sense', where all thoughts are formed, and the soul resides.[36]

Even though his understanding of the pineal gland was wildly misinformed, even according to the scientific knowledge of the day, and was roundly rejected by most of his peers, Descartes' insistence that it was possible to pinpoint the essence of what it is to exist to an exact location within the head marked a turning point in our thinking on the nature of intelligence. Western medicine had long been grounded in the teachings of the Ancient Greeks, who had conceived of a tripartite human soul: the rational soul that resided in the brain, the sensitive soul that found a home in the heart, and the vegetative soul which could usually be found in the stomach or the liver. Their attribution of higher functions like reasoning and memory to the brain chimes with modern knowledge, but less familiar to us is the conviction that other parts of the body were equally active and capable of acting independently from the mind. Galen, philosopher, pioneering anatomist and one of the most acclaimed physicians of the classical age, conceived of the stomach as an autonomous actor capable of registering its own emptiness. He wrote that nature 'has granted to the stomach alone and particularly to the parts of it near its mouth the ability to feel a lack which rouses the animal and stimulates it to seek food'.[37] We are all familiar with the growling demands of an empty belly, but for Galen these served as proof that it possessed its own kind of perspicacity. This understanding of the intelligent gut sustained a broader theory of the body, debated through antiquity, that refused to make any sharp distinction between its physical and mental aspects, allowing for the possibility that thinking might take place at various sites around the body.

By the seventeenth century, however, this vision of a kind of dispersed intelligence was waning in the face of the new anatomical discoveries being made by scientists like Thomas Willis. Often credited as the founder of modern neuroscience, Willis spent his career examining the brain in ever more minute detail, mapping its vast neural networks and complex muscle structures, and eventually publishing his findings in a 1664 treatise titled *Cerebri anatome*. In common with his French predecessor Descartes, Willis conceived of his research as part of a larger enquiry into the site and true nature of the soul. Anatomy was, he proclaimed in the book's opening pages, a method by which it becomes possible to 'unlock the secret places of Man's Mind' and observe 'the living and breathing Chapel of the Deity'.[38] Written in the turbulent aftermath of the English Civil War, which upended religious, social and political certainties, Willis's account effectively reimagined the brain as a creation of the Divine. A committed Anglican, he conceived of scientific observation in deeply religious terms, not as a challenge to God, but as a way of discovering and honouring His presence in the natural world.

Though Willis drew on the work of physicians like Galen – as Professor of Natural Philosophy at the University of Oxford he would have also been required to instruct his students in classical thought – his account of the brain differed in one key respect. While the Ancient Greeks and even early Christians imagined animals to possess something like a soul, Willis and many of his contemporaries viewed it as the exclusive preserve of human beings. According to Descartes, animals were essentially without consciousness, more akin to mechanical objects like clocks than to man. While human capacities like language, creativity and rationality signalled the presence of an immortal soul, the assumed lack of those qualities in animals reduced them to unthinking, unfeeling assemblages. Philosophers like Willis and Descartes proposed that only human brains contained a soul and all that it connoted – emotion, intelligence, immortality – and by doing so they not only refuted the notion that the belly might think, but also restricted the privilege of thought itself to man.

While it is hard to believe that Descartes truly perceived there to be absolutely no difference between the clock sitting on his desk and the pet dog softly snoring beneath it, his dismissal of animals as mindless automata did have material effects, most tangibly in helping to legitimise the practice of vivisection in France. As one physician recalled of his time at the Port-Royal Hospital in eighteenth-century Paris:

> They administered beatings to dogs with perfect indifference, and made fun of those who pitied the creatures

as if they felt pain. They said the animals were clocks; that the cries they emitted when struck were only the noise of a little spring which had been touched, but that the whole body was without feeling.[39]

If animals were simply machines, then what sounded and looked like expressions of pain could be easily dismissed as sentimental impossibilities. Having excluded them from the life of the mind, it became morally permissible to treat animals as if they were not truly living, not truly feeling creatures.

Of course, this philosophical position has never mapped neatly onto real-world encounters between people and animals. Farmers cared deeply for the welfare of their cattle, working animals like horses were praised for their intuition and household pets of all kinds were doted upon by their owners. Yet the everyday experience of living with other species and witnessing their emotional lives – the bored expression of a dog waiting to be taken out for a walk, the squealed delight of pigs in a muddy yard, the distress of a lamb taken from its mother – did not prevent the notion that animals were somehow without consciousness, or at least not conscious in the way humans are, from lodging in some recess of the collective imagination. This is materially evident in the persistence of meat eating or the popularity of leather shoes, but it continues at an even deeper level. According to historian Erica Fudge, most modern philosophers continue to 'assert that the Cartesian human is the only model of the human available' so that the 'human as a being distinct and absolutely separate from the animal is represented as a given'.[40]

In other words, the idea of beast as machine has structured the Western sense of self, informing how we think about questions of individual identity and our broader place in the world. It has also, crucially, helped to establish what actions can be justified and who can be sacrificed in the pursuit of knowledge, particularly scientific knowledge. And yet, despite what the breezy callousness of the vivisectors at the Port-Royal Hospital might suggest, debates over the ethics of animal experimentation have never been comfortably settled.

To return to the Department of Physiology at UCL and to the fate of a small brown terrier: on a brisk February day in 1903 a group of students gathered in a lecture hall for a class with William Bayliss. Strapped to the table in front of him was a dog, alive and semi-conscious, who was to aid in the demonstration of some key physiological principles. That day's lecture concerned the production of saliva, so Bayliss began by slicing open the side of the muzzled dog's neck to expose the glands. Attaching electrodes to the surrounding nerves, he attempted to show that the production of saliva was not dependent on blood pressure. The lecture did not, however, go to plan and after struggling to make his point for some time, Bayliss handed the dog over to his assistant who removed the pancreas before killing it with a knife through the heart. Sat in the audience, making copious notes on everything that occurred, were two interlopers. Founding members of the Swedish Anti-Vivisection Society Leisa Schartau and Louise Lind af Hageby had enrolled at the London School of Medicine for Women to get a closer look at the practice they had committed themselves to eradicating. After attending several demonstrations, the pair presented their findings to Stephen Coleridge, a leading figure in the anti-vivisection movement, who suspected foul play.

Though vivisection was legal in Britain, the practice was highly controversial. While the medical community pointed to the scientific value of experimenting on live animals, to many it appeared unnecessarily cruel and ultimately unjustifiable. Opposition was compounded by the fact that the animals most vivisected were the same creatures – dogs, cats, rabbits – that, in different circumstances, might be considered members of the family. The image of beloved pets dying miserable prolonged deaths in classrooms around the country prompted the Anglo-Irish feminist Frances Power Cobbe to establish the National Anti-Vivisection Society in 1875. Far from a marginal concern or radical cause, the campaign against animal experimentation counted Queen Victoria as one of its supporters, who condemned it as 'horrible, brutal and unchristianlike'.[41] Given that the royal household included several collies, a greyhound, two Skye terriers, a beloved spaniel and a small herd of Pomeranians, it is perhaps unsurprising that the queen felt the practice of vivisection to be an unacceptably 'brutal' one (Figure 3).

The question of what was and was not justifiable in the pursuit of new knowledge began with the 1875 royal commission convened to examine the 'Practice of Subjecting Live Animals to Experiments for Scientific Purposes'. The inquiry heard testimony from multiple experts on both sides of the debate, but the account given by one witness proved particularly influential. Under examination Emanuel Klein, a Croatian-born bacteriologist and

Figure 3: Animals dressed as doctors are about to
dissect a man in an operating theatre, 1910

lecturer in histology at St Bartholomew's Hospital, claimed to have 'no regard at all' for the suffering of animals and admitted to only ever using anaesthetic to avoid being scratched or distracted from his work by cries of pain.[42] His apparent indifference to the suffering of living creatures shocked the commission into recommending a raft of new measures – animals were to be anaesthetised, could only be used once and had to be killed humanely once the experiment was complete – which were designed to check the viciousness of bad apples. For opponents of vivisection, however, Klein's statement only served to confirm what they had already suspected: that doctors, hearts hardened by years of medical school, thought little of inflicting pain on harmless creatures in the name of so-called 'scientific progress'. Such wanton cruelty was, many felt, against the spirit of an age that had seen a major shift in attitudes towards the rights of non-human creatures. Alongside the introduction of important pieces of legislation like the 1835 Cruelty to Animals Act and the founding of charitable organisations like the Royal Society for the Prevention of Cruelty to Animals, popular cultural depictions of animals encouraged a view of them as virtuous, loving and deserving of kindness. The practice of vivisection, then, contravened an emerging set of social norms around the value of non-human life, but it also suggested that patients should be wary of placing too much trust in their doctors. After all, was it wise to rely on the care of a profession who appeared entirely unmoved by the pain of living things?

It was with questions like this in mind that, having reviewed the account of the UCL dissection, which suggested that not only had the dog not been properly anaesthetised, but also that the audience of students had apparently found its cries of pain to be a source of much hilarity, Coleridge decided to go public with his concerns. After an article in the *Daily News* printed the accusations and questions were raised in the House of Commons, Bayliss was forced to file a lawsuit for libel. The case was heard at the Old Bailey in November 1903 and even though Coleridge lost the case, the trial was ultimately a victory of the anti-vivisectionist cause. Reports in the popular press replayed the grisly details of the experiment, membership to the National Anti-Vivisection Society increased and money was even raised to formally memorialise the dog. Erected on the Latchmere Recreation Ground in Battersea in 1906, the statue (Figure 4) was inscribed with the following call to action:

In memory of the Brown Terrier Dog Done to
Death in the Laboratories of University College in
February 1903 after having endured Vivisections
extending over more than two months and
having been handed over from one Vivisector
to another till death came to his Release.

Also in memory of the 232 dogs vivisected in
the same place during the years 1902–3.

Men and Women of England
How Long shall these things be?

Provoked by this epitaph, medical students vandalised

Figure 4: The Brown Dog statue

the statue, riots broke out across London and eventually it was removed by the local council, who were tired of the controversy and unwilling to keep it under twenty-four-hour police guard. The tumult, which saw groups of male students clash with suffragettes and young men from the local area, underscored the complex class and gender dynamics at play in the vivisection debate. For feminist thinkers like Frances Power Cobbe there existed a clear link between the cruelty meted out in the laboratory and the violence directed at women, and working-class campaigners forged cross-species solidarity based on having had their own bodies similarly exploited and misused by those in power. While the medical community viewed vivisection as a contributor to the collective good, furthering our understanding of the body and its ailments, for their opponents the practice revealed only the fact that some bodies were worth less to society than others.

The question of bodily worth, how to quantify it and how to weigh the needs of some bodies over others, had also preoccupied the 1875 royal commission into vivisection. Asked to wrangle with these tricky ethical questions, witnesses to the commission looked to intelligence as the most useful measure of value available. Proposing a tightening of regulations, one doctor argued that animals with greater 'reasoning or thinking power' were likely to suffer more pain and on those grounds proposed a ban on the use of monkeys, a prominent anti-vivisectionist condemned the misuse of a particularly 'intelligent' dog in physiological experiments, while those speaking in defence of the practice justified it on the grounds that animals lack the necessary intelligence to anticipate or remember pain.[43] For both sides of the debate, animal behaviour could be measured in relation to human capacities, so that those who most closely resemble us were perceived as thinking creatures and thus, for some, more deserving of kindness, respect and even rights. Implicit in this was an understanding of intelligence as primarily the property of the human brain, a top-down model that seemed to confirm the natural superiority of mankind. And yet part of what the work of William Bayliss and Ernest Starling revealed – research that was undertaken on the unwilling bodies of countless dogs – was that intelligence might consist of rather more than what is contained by the hard walls of the skull. Their discovery of secretin, a chemical messenger produced by the intestine and released into the bloodstream, suggested that the gut was not – as had been previously assumed – tightly controlled by the central nervous system, but was in fact an autonomous and powerful agent. Taken seriously, the thinking gut could challenge dominant understandings of what intelligence is, who possesses it and the extraordinary value that comes to be accrued to it.

Today, as the climate crisis gathers pace, scientists from a wide range of fields, working alongside philosophers, writers, artists and activists, have argued that recognising different kinds of animal intelligence might help us to rethink our position in the world. Take, for instance, the octopus. A member of the cephalopod class, this spineless eight-armed creature has recently garnered some celebrity off the back of videos circulating across social media that record this natural escape artist: large bodies squeeze through impossibly small cracks, tentacles deftly prise open doors and unscrew lids, laboratory floors are smoothly slithered across, and drainpipes elegantly descended. More than simply charming, these antics also provide a glimpse of their remarkable intelligence. Octopuses use tools, engage in problem solving, learn by observing their environment and communicate using a remarkably diverse range of signals. They are capable, in other words, of what are known as high-order cognitive behaviours. And yet the octopus's brain does not resemble – even remotely – that of humans. Each of their eight arms is loaded with neurons and each can make decisions independently from the central brain, which is located not in the head, but looped around the oesophagus. This means that everything they ingest must first pass through the brain, an anatomical quirk that positions digestion at the heart of cognition in a way that upends the hierarchy of the body and points to a radically different way of envisioning what intelligence might look like. Even organisms invisible to the naked eye have been shown to have cognitive potential. Recent research into how bacteria in the body, and especially in the gut, fight viruses, found that they copied parts of their DNA to incorporate into their own, so that the next time the same virus attacks, the bacteria can help the body to mount a sustained defence. What this demonstrates is that bacteria seem to be capable of processing and interpreting information, then connecting it with contextual meaning, engaging in what the scientists describe as 'non-conscious cognition'.[44] All of which suggests that intelligence may be more multiple, more dispersed, more bodily, more creaturely and less human than we like to assume.

2

Dangerous Passions

In Elvis Presley's 1957 homage to heartbroken holidays, 'Blue Christmas', colour and emotion stand in for one another: against the cheerful red, white and green of the festive season, he pits the blue, blue, blue, blue of sadness. As listeners we make this connection without thinking, but it is worth asking how we came to 'feel blue'. Why is the colour blue, at least in English, so associated with low and melancholic moods? Explanations vary from Greek mythology to fourteenth-century poetry, from the use of blue indigo in West African funeral rites to the naval custom of flying a blue flag to mark the death of the ship's captain. The *Oxford Dictionary of Phrase and Fable* offers yet another possibility in the form of small blue-skinned devils. According to this account, from around the middle of the eighteenth century, the term 'blue devils' was used to describe states of depression and nervous collapse.[45] In a letter dated 1819, the Romantic poet John Keats admitted to having lately struggled with 'the blue-devils'; the art historian Anna Brownell Jameson knowingly referred to her semi-autobiographical account of travelling through Italy in the 1820s as the 'Diary of a Blue Devil'; and in Lord Byron's epic *Don Juan* (1819) even the notorious womaniser

'Hath got blue devils for his morning mirrors'.[46] These mischievous little demons could be summoned by over-work or misguided romantic dalliances, but they might also arise from the functions of the body. In his *Essay on Indigestion* (1827) James Johnson, renowned surgeon and physician to William IV, described the blue devils as the result of 'poisonous secretions' lurking in the bowel, 'the nerves of which are so numerous and the sympathies so extensive, there is induced a state of mental despondency and perturbation'.[47] Elsewhere, the connection between digestive trouble and low mood was sustained in George Cruikshank's satirical cartoon on the folly of overindul-gence, where blue devils are among the sprites wielding instruments of torture, while miniature servants goad the sufferer with plates of rich food (Figure 5). That 'feeling blue' might originate in the rumblings of the discontented belly suggests that the relationship between the digestive system and the mind was not always a friendly one. Which is not all that surprising. There is a disconnect between the lofty aspirations of the mind and the unpalatable work of the gut, between the ambitions of an organ that thinks, reasons and creates and one that consumes, assimilates and defecates.

Today we tend to imagine the gut–brain axis in affirming terms, as a nurturing, restorative and mutually beneficial connection to be carefully cultivated. The recent fashion for fermented foods, which brought glamour to pickled cabbage for the first time, was driven by research into the role that lively intestinal bacteria might play in sustaining mental wellbeing. Some scientists have isolated the vagus nerve as a key pathway from intestine to hypothalamus,

Figure 5: A man suffering from indigestion, as suggested by the little characters and demons tormenting him. Etching by George Cruikshank, after Alfred Crowquill, 1835

while others have studied neuroendocrine signalling for insight into the role of the microbiome in the production of dopamine and serotonin. Experts in neurogastroenterology have also explored the potential downsides of all this gut–brain communication, pointing to the role that depression and anxiety play in exacerbating the physical symptoms associated with conditions like IBS. Even then, such complaints are typically framed as the result of a rocky patch in a long-term relationship, in which love can be rekindled by simply paying more careful attention to the needs of each partner.

The marriage of stomach and mind has a more troubled history than our current rosy view of that union might suggest. Ancient philosophers like Plato and Aeschylus regarded the bowels as the site of base passions – anger, jealousy, desire – and tasked the higher faculties of the brain and heart with keeping the guts in check. Writing in 1497, Alessandro Benedetti, the surgeon general of the Venetian army, similarly disparaged the bowels as 'relegated farther away from the site of reason' and 'fenced off' by the 'diaphragm in order not to disturb the rational part of the mind with its importunity'.[48] In common with many anatomists of the period, Benedetti regarded the intestines as the site of not only body waste, but also spiritual impurities and violent emotions. Even into the nineteenth century, the gut was often characterised as a source of bad temper and foul mood. The stomach, according to one prominent Victorian physician, was 'a strangely wicked and ungrateful organ [...] implacably opposed to man's progress and comfort'.[49] What emerges across time, then, is a rather ambiguous picture of the congress of stomach and mind: are we wise to follow our gut feelings? Or might we be led astray by the gurgling of an untrustworthy organ? This chapter offers up a rather more acrimonious reading of this relationship, one defined by discord, distrust and even demonic possession. Following in the impish footsteps of those wicked blue devils, it takes a journey through the darker recesses of history to uncover the strange, magical and sometimes malignant forces that have found a home in the gut.

Go to the wellness section of any bookshop and you will encounter numerous titles devoted to gut-healthy eating. Bestsellers like Eve Kalinik's *Happy Gut, Happy Mind* (2020) and Naomi Devlin's *Food for a Happy Gut* (2017) combine nutritional advice with recipes formulated to ease digestion, repopulate the microbiome and propagate good bacteria. By far the boldest claim made by this genre of diet book is that eating the right kinds of foods has a direct effect on mental wellbeing. According to this new generation of self-help gurus, emotional health can be achieved by consuming a combination of wholegrains, nuts and organic vegetables, because these nutrient-dense foods help to cultivate intestinal bacteria that – through a complexity of nerves, hormones and cross-organ telegraphing – lift the mood of the eater. As improving as all this sounds, though, there is surely more to the relationship between what we eat and how we feel than only the pursuit of happiness. Food can provoke all kinds of emotions, both positive and negative: the warm glow of nostalgia emitting from a favourite childhood treat, frustration at the inadequacy of an overpriced sandwich bought in haste, joy at the simple pleasure of a ripe peach, dread as your dinner host places a long-loathed dish on the table. There is, moreover, no simple correlation between consuming foods considered nutritionally sound and feeling happy. A lunch of cheeseburger, fries and milkshake is as likely to provoke delight in the eater as a well-balanced meal is to inspire profound despondency; additionally, the idea of gut-friendly eating rests on certain assumptions about the nature of the organ that do not hold true across time.

The link between gut health and emotional wellbeing relies on a distinctly modern understanding of consumption, which historian of science Steven Shapin has described as 'analytic' in character. Visualised by the nutritional breakdown on the back of supermarket packages, we have been taught not only to think of food in terms of its chemical composition, as a conglomerate of carbohydrates, proteins, fats, minerals and vitamins, but also to view ourselves as somehow similarly assembled. As Shapin has it, 'You are a bag of chemicals, organized into physiological systems; eat the right chemicals and you will enjoy good health; eat the wrong ones, and you will suffer disease and shortened life.'[50] This model relies on an honest bargain with the gut – feed it well and you will enjoy fine health – but history offers up plenty of instances in which the organ has been charged with duplicity and malevolence. Accusations that are made even more troublesome by the influence it has been rumoured to have over the mind, which has long been envisioned in ways that go far beyond the reliable production of happy hormones. Indeed, beyond the many roles it plays in our social and cultural world, at different times throughout history the food we eat and the process of digestion has been held responsible for shaping our emotional lives and unconscious experiences in ways that far exceed the neat system of inputs and outputs described by Shapin.

The cheese dream is a case in point. A 2005 study conducted by the British Cheese Board investigated the popular belief that eating cheese before bed results in weird dreams. Having surveyed two hundred volunteers, the study found that certain types of cheese seemed to induce different kinds of dreams: Stilton led to bizarre dreams, Red Leicester provoked dreams of the past and

Cheddar encouraged celebrities to make a nocturnal appearance. Designed as a publicity stunt, this survey was particularly effective because it spoke to the persistent belief that digestion exercises some influence over the dreaming mind. In the early nineteenth century French physicians blamed 'pain in the epigastric region' for the night terrors that afflicted their patients; into the 1940s *Webster's Dictionary* continued to define the nightmare as 'a condition brought on in sleep, commonly by digestive or nervous disorders and characterised by a sense of extreme discomfort or by frightful dreams'; and more recently, research published in the *American Journal of Gastroenterology* has explored the possible relation of bad dreams to bad bowels.[51] Such connections point to a long-standing appreciation of the interaction between gut and mind, but they also draw attention to the dream as a peculiarly complex phenomenon. We have always dreamed, but over time the language we have used to describe our nocturnal visions and the meanings we have chosen to ascribe to them have changed dramatically. In the ancient world dreams were visitations, bringing glimpses of the future or important messages; for early Christian thinkers the dream was a means of excoriating self-revelation essential to keeping believers on the path to salvation; while for readers of dream books, which were hugely popular in Victorian Britain, they were puzzles filled with intriguing symbols waiting to be decoded.

In the medieval period dreams were often thought to be direct communications from God, but such claims were open to interrogation. Dream theorists were tasked with distinguishing between visionary dreams – possibly divinely inspired, possibly demonically motivated – and the mundane fancies of the everyday. In this process timing was everything.[52] Dreams that occurred in the early morning were understood to be more likely to be of a pure or even prophetic nature because they took place long after the evening meal had been fully digested; undisturbed by the lower processes of the body, the sleeping soul was thought to be open to higher inspiration. This relationship between digestion and sleep was also, as historian Sasha Handley has uncovered, a prominent feature of early literature on health. Readers of books like Thomas Elyot's *Castel of Health* (1539) and Thomas Cogan's *Haven of Health* (1584) were instructed to base the number of hours they slept on the kind of food consumed and the strength of their digestive system. So, a young person, in good health, who ate moderately required less sleep than older people, gluttons, or pregnant women, whose digestion was slowed by the weight of the foetus on the stomach.[53] Failure to take the gut into account would result in disrupted slumber, night terrors and sleepwalking.

Even as dreams became an object of scientific study, attracting the attention of physicians, physiologists and psychiatrists, the stomach retained some of its influence over the sleeping mind. In his *Philosophy of Sleep* (1827), for instance, the Scottish physician Robert Macnish devoted several chapters to a detailed consideration of how digestion shapes sleep. Eating a heavy meal just before bed risked a poor night's rest, a bout of indigestion was usually accompanied by nightmares and a disordered 'liver will often produce dreams' of a peculiar kind.[54] In the early twentieth century, this physiological understanding of dreams as the result of forces and processes within the body proved remarkably resilient in the face of increasingly sophisticated psychological theories. Hailed by Sigmund Freud as the 'royal road to a knowledge of the unconscious activities of the mind', the careful study of dreams was key to the emerging discipline of psychoanalysis.[55] A therapeutic technique that promised to unearth deep desires, forgotten memories and buried traumas, dream interpretation was based on a model of mind that seemed – at first glance – to have very little to do with the rumblings of the gut. However, a letter from Freud to his friend and collaborator Wilhelm Fliess complicates this picture. On 31 October 1897, he recounted that his daughter Anna – then nearly two years old – had called out in her sleep for 'Stwarberries, high berries, scwambled eggs, pudding!' That morning she had been sick, and on her nanny's orders, not been permitted to eat anything for the rest of the day. By calling out 'a whole menu in her sleep', little Anna was, by her father's account, engaging in the simplest kind of wish-fulfilment, dreaming of the foods she desired and had been denied.[56] Sometimes the royal road to the unconscious takes a detour through the stomach.

Since 2015, the website cheesedeutung.tumblr.com has been inviting submissions to its archive of fromage-induced dreams. Contributors are asked to specify the type, price and amount of cheese consumed along with a description of the resulting dream. After sampling 'three varieties' of Manchego comes a dream of cancelled flights and an airport fire; smoked Cheddar at 9 p.m. produces a lengthy somnambulant car chase through the streets of an unknown city; a four-cheese pizza just before bed and Fidel Castro is mayor of New York; and strong goat's cheese leads to a muddle of murder, sadistic husbands, turquoise silk dresses and snowy motorways.[57] According to folklorist Caroline Oates, who has investigated the history of cheese and nightmares, this association is particularly prevalent in the United States and Britain, but far less widely reported elsewhere in Europe. That the French are unlikely to be bothered by Camembert and Italians go unmolested by late-night Taleggio suggests that social attitudes towards certain foods may be as much to blame for strange dreams as the cheese itself.

As well as being culturally inflected, cheese dreams are also historically specific. Their heyday came in the opening decades of the twentieth century, where cheese-engendered reveries were a subject of fascination for satirical cartoonists and early filmmakers. On 10 September 1904, the New York-based *Evening Telegram* featured a comic strip titled *Dream of the Rarebit Fiend* that detailed the nocturnal misadventures that resulted from overindulgence in the Welsh delicacy. Penned by Winsor McCay and printed in some form until 1925, the cartoon depicted dreams that ranged from the fantastical to the disturbing, but which usually ended with that week's 'fiend' swearing never to eat cheese before bed again. One strip depicts the hapless dreamer being buried alive, in another a suitor befuddled by the object of his affections sees her disintegrate into hundreds of puzzle pieces and elsewhere an evening bath is disrupted when a huge hungry hippo surfaces in the tub. Credited with having prepared the popular American imagination for the radical insights of psychoanalysis

– Freud gave a series of lectures in New York in 1909 and the first English translation of *The Interpretation of Dreams* appeared in 1913 – *Dream of the Rarebit Fiend* delved into the nation's buried anxieties, secret desires and dark wishes.[58] But it also helped to sustain the connection between mind and gut, the comic potential of which was taken up by some of the earliest filmmakers. Along with an adaptation of *Dream of a Rarebit Fiend* (1906), silent films like *The Glutton's Nightmare* (1901), *Baron Munchausen's Dream* (1911) and *A Vegetarian's Dream* (1913) all feature protagonists who overindulge at dinner and spend a fretful night as penance. Why was the United States such a hotbed of gastric nightmares? One answer lies with rarebit. References to the dish date back to eighteenth-century Britain and it became popular across the Atlantic in the early twentieth century. The cheesy delight was a favourite of late-night diners: theatregoers, city workers finally leaving the office and drinkers looking to soak up the excesses of the tavern with a cheap, filling snack. Because of this, rarebit became synonymous with the pleasures and perils of the growing city, signifying both the thrilling whirl and untenable strain of modern life.[59]

The world depicted in Winsor McCay's long-running comic strip is a darkened urban landscape of tall buildings, towering steeples and dim street lamps where cheese-mad dreamers wrestle with the stresses of their day. The central conceit of this cartoon and films from the period like *Baron Munchausen's Dream* – in which the drunken and overfed protagonist is assailed by frog-like monsters, giant grasshoppers, dragons and a hideous spider woman – is that the workings of the digestive system can generate visions so vivid that they might be mistaken for reality. The rumblings of the gut have often been called upon to explain strange dreams or seemingly supernatural experiences. When in *A Christmas Carol* (1843), Ebenezer Scrooge blamed one of his spectral visions on 'an undigested bit of beef, a blot of mustard, a crumb of cheese, a fragment of an underdone potato', he did so in line with a century-long understanding of ghost sightings that attributed all things supernatural to trickery, illusion or delusion.[60] This way of thinking about ghosts was, as historian Shane McCorristine has argued, a product of the intellectual revolutions of the eighteenth century. He writes that while the 'ghost had a purpose and a place in medieval society', the Enlightenment relocated 'the ghost from the external, objective and theologically structured world to the internal, subjective and psychologically haunted world of personal experience'.[61] In other words, the rise of rational secularism over the course of the eighteenth century redefined supernatural experience as pathological evidence: phantasmal visions were produced by a deluded eye, spectral sightings were warnings of a disordered mind and spooky dreams only the work of an upset stomach.

This debunking tradition can be traced back as far as the sixteenth century and specifically to the publication of Reginald Scot's *The Discoverie of Witchcraft* (1584). Published as witch trials were taking place in towns and villages across England, this controversial treatise characterised the belief in witchcraft and sorcery as fundamentally irrational. Building his argument, Scot began with the widely held belief that nightmares were the work of black magic. There was, he wrote, a perfectly rational explanation: bad dreams were in fact 'ingendered of a thicke vapor proceeding from the cruditie and rawness in the stomach, which ascending up into the head oppresseth the braine, in so much as manie are much infeebled thereby as being nightlie haunted therewith'.[62] Fearful visions and night terrors were the result, then, not of malevolent spirits or vengeful crones, but of half-digested food sat fermenting in the gut; and the solution lay neither in prayer nor in burning innocent women, but in proper mastication and careful diet.

Pioneered by sceptics like Scot, whose book was purportedly burned once the notorious witch-hunter James I ascended to the throne in 1603, rational theories of the supernatural would eventually prevail. But this is not the whole story. After all, the three ghosts Scrooge encounters that fateful Christmas Eve are not merely the products of an 'undigested bit of beef'; they are real or, at the very least, his encounter with them produces real effects in the real world. Alongside the explanations offered by science, mythical, folkloric and magical accounts have persisted into the modern world. This duality is captured in an engraving of a painting by Henry Fuseli that hung in Freud's apartment in Vienna (Figure 6). First displayed in 1781 to popular acclaim, *The Nightmare* depicts a sleeping woman surmounted by a demonic figure and observed by a white-eyed horse who peeks through red velvet drapes. Dealing in the long-established symbolism of dreams – the incubus who haunts female sleepers and the 'mare' that brings frightful fancies – the painting has often been read as a depiction of the darker recesses of the unconscious mind. What the image also reveals, however, is the remarkable degree to which dreams were shaped by folk

Figure 6: Henry Fuseli, *The Nightmare* (1781)

beliefs and supernatural lore. To fall asleep is to make yourself vulnerable to the machinations of all kinds of malevolent creatures: in Germany the 'mare' might turn you into a horse to be ridden by witches, in Scotland bad faeries could lure you to their nightly balls and in Italy a cat-like creature called the *pandafeche* might strike you down with paralysis. Even seemingly natural causes could have supernatural origins. Poor digestion might appear to offer an unassailably rational explanation for bad dreams, but it still left open the possibility that supernatural forces were at work in the gut. Witches, for example, were often charged with having brought on indigestion to induce a nightmare and the powers of the 'Evil Eye' could obstruct swallowing. As Caroline Oates has uncovered, in early modern Europe 'both nightmares and indigestion could be attributed to spirit assault', so that gastric upset might be due to the consumption of spoiled food, but it might just as easily be interpreted as a sign of malevolent forces at work.[63]

The stomach has been characterised as an organ that is especially amenable to demonic intervention. In the third century, the Syrian philosopher Porphyry of Tyre warned his reader against eating meat on the grounds that demonic entities lurked in the flesh of dead animals waiting for the opportunity to enter the human body and sow corruption within. His five-part treatise *On Abstinence from Killing Animals*, which was purportedly written to convince his friend Firmus Castricius to take up the vegetarian diet, stresses the spiritual jeopardy that carnivores place themselves in every time they sit down to eat. Evil *'daimones'*, he wrote, 'rejoice in meat on which their pneumatic part grows fat, for it lives on vapours and exhalations, in a complex fashion and from complex sources, and it draws power from the smoke that rises from blood and flesh'.[64] Once imbibed, the hungry demons harness the digestive system of their unwitting host, causing noxious gasses to build that would result in bloated abdomens, belching and flatulence. Only by carefully policing the appetite and abstaining from the fleshier temptations of the table could one avoid welcoming disruptive spirits into the body.

Later, during the witch trials that took place across Europe from the fourteenth century onwards, questions of consumption and digestion were taken up in the detection and prosecution of those accused of practising black magic. Food was both a tool used in the practice of magic – from the herbs gathered for spells to the unholy meals shared on the sabbath – and itself a target of witchcraft. As historian Chris Kissane has explored, failed harvests were often attributed to the evil machinations of local witches and poisoning was thought to be one of the ways that witches enacted revenge on those who had wronged them. Milk, according to Kissane, was a source of particular concern: he writes that 'demonology literature is full of stories of women (and it was mostly women) who could magically "steal" milk, spoil it, make it impossible to churn or cause the milking animal to "dry up"'.[65] Part of what the discourse around so-called 'milk magic' reveals is anxieties over the power that women were able to exercise over consumption. Having been charged with managing the kitchen, tending the garden and looking after any animals, wives and daughters effectively controlled what was drunk and eaten in the home. That so many accusations of witchcraft centred on the poisoning or spoiling of food suggests an unease with this arrangement and suspicion that wronged, vengeful or simply malicious women might be tempted to act out their frustrations on the stomachs of their menfolk.

The association of milk with magic also speaks, perhaps most obviously, to women's childbearing capacity. Though historians have quibbled over the number of actual prosecutions secured, it is certainly the case that midwifery and witchcraft were often conflated with one another within early modern culture.[66] Midwives were charged with possessing devilish knowledge of herbs and natural remedies, with replacing children with other-worldly changelings and, most often, with bringing about miscarriage and stillbirth. In *The Malleus Maleficarum*, a guide to witch-hunting published in the fifteenth century, its author, a clergyman named Heinrich Kramer, claimed that 'no one does more harm to the Catholic Church than midwives', who – when they are not killing and eating newborn babies – are likely to be found 'raising them up in the air, offering them to devils'.[67] The widespread distrust of midwives and the horrible acts they were accused of committing speaks to the immense authority they wielded within their local communities. Possessing intimate knowledge of the mysteries of procreation, fertility and birth, their seeming power over matters of life and death, in a patriarchal society not generally amenable to displays of female mastery, exposed them to charges of malfeasance.

Before the invention of the pregnancy test in the 1920s and the development of ultrasound technology in the 1950s, pregnancy was a rather ambiguous state of being. Until the quickening, when it becomes possible to feel the baby move in the uterus, it was unclear if conception had occurred and after that impossible to tell whether it could be carried to term. Pregnancy was made more mysterious by terminological and conceptual slippages that collapsed distinctions between gestation and digestion. In her study of divine and demonic possession in the Middle Ages, historian Nancy Caciola draws attention to the overlaps between womb and gut in this period. The Latin term *venter* was used to refer to the womb, the stomach or the digestive tract, and Caciola suggests that this overlapping definition indicates a 'pre-existing cultural assimilation between these organs and their processes'.[68] Swollen bellies were synonymous with not only pregnancy and gastric trouble, but also demonic possession. As historian Boyd Brogan explains, 'in possessed women, and sometimes men, this swollen belly was often compared to pregnancy. A demon was a physical presence within the body, and it made sense for it to take up residence in the space that could stretch to accommodate a child.'[69] It was not, however, always clear exactly where the demon lay in the body, and the bowels were another favoured hiding spot. Demons were thought to be drawn to excrement, so that, according to the sixteenth-century French witch-finder Nicholas Rémy, 'very often he [the devil] has his dwelling in those parts which, like the bilge of a ship, receive the filth and excrements of the body'.[70] A site of dirt and corruption, located far from the higher faculties of heart and mind, the entrails were where dark forces – like a child slowly growing in the womb – made their home in the body.

The gut and womb are alike because they are both spaces in the body where the boundary between interior and exterior becomes permeable, inner chambers where self and non-self get muddled: the baby that is you but also not you, or food from the outside world that is now part of what makes up the inside world of you. As the blue devils that foisted belly aches and melancholic feelings on their unhappy victims reveal, with openness comes vulnerability. The image of the stomach as a remote and lawless outpost where demonic forces might gain a foothold reflects a way of thinking about that organ quite different from the way it is conceptualised in medicine and popular culture today. Though there is, undoubtably, something troubling in the idea that forces outside of your control – whether they be masked demons or tiny blue devils – might take up residence in the viscera, bringing with them bad dreams and negative feelings, perhaps possession also has some upsides. Would it not be nice, sometimes, to be able to blame a bad mood on a supernatural entity? Instead of interrogating hidden motivations, examining patterns of behaviour or re-treading past traumas with a therapist, one could hold an outside force responsible, have an exorcist banish them and all the problems they bring. Perhaps there is a similar impulse at work in our current enthusiasm for the idea of a gut–brain connection. Locating emotions in the body, in the serotonin produced by gut microbiota, say, or in the telegraphing of the vagus nerve, can help make the unintelligible – a flash of frustration, a wave of sadness that seems to come from nowhere, a gloomy mood that will not shift – seem intelligible. Not only intelligible but treatable through the adoption of a better diet, the cultivation of more multiflorous gut flora and so on. Yet history suggests that the gut may not be the most reliable organ on which to base one's sense of mental wellbeing. The belly has more often been viewed as a cause of emotional turmoil, resistant to discipline and open to the machinations of malevolent forces.

Even when no demonic force was involved, the gut has often been held responsible for throwing the body and the mind into chaos. In the same moment as demonologists were seeking evidence of the devil in the bloated stomachs of accused witches, early modern doctors were attributing diseases of the brain to the rumblings of the stomach. Through a process commonly known as 'sympathy', toxins from one body part could cause symptoms to emerge elsewhere, so that winds and fumes emitting from the belly could rise up and bring about everything from melancholy to headaches to seizures. For example, in one of the first treatises to debunk the demonic theory of nightmares, *Hidayat al-Muta'allemin fi al-Tibb* (*Learner's Guide to Medicine*), the tenth-century Persian polymath Al-Akhawayni Bokhari attributed them instead to vapours of phlegm ascending from the stomach and suffocating the brain during sleep. In place of exorcism, he prescribed bloodletting and 'thinning of the diet', insisting that poor gastric health was to blame for night terrors. Leaving impish blue devils and witchery behind, the following chapter explores the question of diet. Faced with the devious misrule of the possessed, ungovernable, irretractable, irritable gut, throughout history we have turned to diet to wrest back control of our mental and physical wellbeing. With, as we will see, varying degrees of success.

3

Taming the Belly

If you are ever invited to take tea at Buckingham Palace, it is very important that you remember that the handle of your cup should always be turned to a right angle of ninety degrees. Also, please do not embarrass yourself by pouring milk into the cup before the tea or forgetting that your saucer should remain on the table while you sip – unless you are moving around the room, of course, in which case it ought to be held by the left hand while the right gently lifts the cup. But as this is afternoon tea you will be more likely seated on something overstuffed and antique, so careful not to spill! Accompanying delicacies must always be eaten in the correct order – sandwich, scone, cake; one cannot skip blithely back to savoury after indulging in sweet – and whatever you do, don't slurp!

Tea drinking at the palace is a serious business. So much so that before joining her future in-laws for an afternoon cuppa, Meghan Markle reportedly undertook intensive etiquette training to avoid making a disastrous faux pas like over-stirring the cup. Such efforts were perhaps intended as protection against the tabloid view of her as an 'uncouth', 'unsuitable' match for a member of the royal family. After all, knowing how to behave at the table – what fork to use for what, how to compliment your host, what topics to discuss and those to avoid – is the mark of a true insider, a clear sign that you belong where others do not. This rule applies not only to the aristocracy. While the king might allow you to pick up a small cucumber sandwich with your left hand, in Ethiopia it is considered unclean to eat with anything but your right hand. Slurping in China is an accepted sign of enjoyment and a compliment to the chef, while loud chewing noises are considered the height of rudeness in Brazil. Countries as diverse as South Korea and Turkey share customs – in both cultures the eldest diner begins and ends the meal – while bordering nations like France and Germany have very different takes on proper dining conduct – in the former splitting the bill is rather gauche, while in the latter it is so expected that waiters will usually enquire 'zusammen oder getrennt?' ('together or separately?'). From palaces and expensive restaurants to well-scrubbed kitchens around the world, table manners help to create shared cultures and shape group identities by distinguishing local from tourist, insider from outsider, polite from impolite.

Rules and expectations change from place to place, but they also shift dramatically across time. Throughout the first part of the twentieth century American readers seeking guidance on good manners turned to the best-selling Etiquette in Society, in Business, in Politics, and at Home (1922). Written by Emily Post, a New York socialite who became a respected authority on correct conduct, the book steers the uninitiated through the complexities of everyday life in the upper echelons of American society, from how to receive guests to the potential pitfalls of a garden party and how to negotiate a 'supper that is continuous' or what we might now refer to as a 'buffet'. About formal dinners, Post warns that these are 'not for the novice to attempt'. Such meals are 'the *magna cum laude* honours' of hosting, requiring well-balanced seating plans, highly polished silverware, tempting multi-course menus, an army of well-trained servants and a butler on hand to announce the arrival of particularly distinguished guests.[71] That even breakfast entails elaborate table settings, special plates for butter and bread cut to a narrowly specified width, makes *Etiquette* a rather impractical guide for anyone lacking a sizeable private income and household staff, and it is telling that in its modern incarnation, *Emily Post's Etiquette: Manners for a New World* (2011), expectations have been significantly scaled back. What remains, however, is a preoccupation with table manners as not only a mark of refinement, but also a prerequisite of civilised society.

Opening a lengthy chapter on the subject, Peggy Post – great-granddaughter-in-law and director of the Emily Post Institute – defines the role of table manners as a means of distracting from the unsightliness of alimentation. The problem with eating, she writes, is that it is at once a 'gross activity' involving the mashing, pulping and swallowing of food, and a profoundly 'social activity' that is essential to maintaining the fabric of our shared world. Such oppositions can only be reconciled by the careful cultivation of customs, rules and codes of behaviour that, as the original Post put it, help to 'avoid ugliness' at the table.[72] Taking a tour through hundreds of years of advice on etiquette and diet, this chapter explores the complicated and conflicted politics of the dinner table. Eating is not only a way of fuelling the body, after all, it is also a potent cultural ritual imbued with meanings that reach far beyond the plate. As the anxious advice doled out by the Emily Post Institute reveals, the life of the table is wrapped up in intricate ways with how we feel about our bodies and the messy business of digestion.

Where the previous chapter gloried in the disruptive and demonic influence of the gut, this one turns to the question of how to manage this notoriously cantankerous organ. Diet offers the most obvious answer, with centuries of dietary guidance sold on the promise that by eating certain foods and avoiding others it is possible to claim dominion over the gastric region. Significantly, much of this advice also addresses itself, in different ways, to the emotions. From theological tracts that prescribed plain meals as aids to spiritual piety to popular health literature that linked regimen to individual temperament and the medical thinkers who were so convinced that passion led to digestive upset that they cautioned patients to avoid extreme emotions at dinner, the link between the gut and the mind has often been articulated through the language of diet. Whether this involved altering diet to improve mood or altering mood to improve digestion, the muddying of eating and feeling has – as this chapter will argue – profound implications for how we structure and govern our social worlds.

For Victorian moralists who valorised the cosy domestic sphere as a haven from the compromising temptations of the public realm, what happened in ordinary kitchens had the potential to transform the world. Written during the ascendancy of the British Empire, Isabella Beeton's best-selling *Book of Household Management* (1861) positioned the respectable middle-class home at the vanguard of the colonial project. There existed, according to Beeton, an important distinction between eating and dining. While all creatures eat, it is 'not a dinner at which sits the aboriginal Australian, who gnaws his bone half bare and then flings it behind to his squaw', rather 'dining is the privilege of civilisation'.[73] Imagining cooking as an alchemic process by which nature is transformed into culture, Beeton charged her reader with maintaining the boundary between 'civilised' and 'primitive'. Envisaging a culinary evolutionary ladder, where the 'rank which people occupy in the grand scale may be measured by their way of taking their meals', particular dining practices come to signify the superiority of one nation over others.[74] Behind this smug self-assurance lies a deep fear of the forces that manners are imagined as keeping under control: if the diligent housewife was to leave tablecloths un-ironed, forget to polish the cake forks or forgo forks all together, would society as we know it crumble? What restless energies are being held in check by careful deportment?

According to the German sociologist Norbert Elias, etiquette is a form of emotional management. In *The History of Manners*, the first book of his two-volume study *The Civilizing Process* (1939), Elias characterised the Middle Ages as a period that witnessed, alongside the emergence of the nation state and the burgeoning modern world, ever greater constraints placed on social behaviours and the expression of feelings, especially violent emotions like anger or lust. One of the ways that this was achieved was through the popularisation of conduct literature that instructed the reader on the rudiments of civility and imposed, as he put it, ever 'stronger restrictions to certain impulses'.[75] Surveying etiquette books from the fourteenth century onwards, Elias noted how table manners transformed over time from the communal rusticity of the early medieval feast – where dishes were shared and fingers were the utensil of choice – to the elegant refinement of the later seventeenth century that was characterised by a complexity of cutlery and glassware, along with a studied aversion to anything touched by the lips of another. Growing ever more complex and convoluted over time, manners at the table, from what spoon to use to the intricacies of polite conversation, emerged as the gestural language of the ruling classes, one that defined those in power as 'civilised' and those excluded from power as 'uncivilised'. According to Charles Vyse, an eighteenth-century authority on etiquette, 'Nothing shews the difference between a young gentleman and a vulgar boy so much as the behaviour in eating.'[76] More than a way to fuel the body, alimentation came to mark and justify an individual's position in the world.

Table manners maintained not only the social pecking order, but also the natural hierarchy of the body. Codes of decorum around dining serve to both ritualise and regulate consumption: think of the difference between eating in a restaurant and eating at home, between a shared meal with friends and one consumed alone in front of the television. For food historian Ken Albala, it is no surprise that the emergence of the modern state through the sixteenth and seventeenth centuries was accompanied by an intensification of attempts to control and rationalise eating behaviour. Early works on dietary regime considered, Albala writes, 'the government of the body as analogous to that of states' and therefore 'it is no coincidence that the words *regime* and *regimen* share a common root in the Latin *regere*, to rule'.[77] The well-managed body was alike to the well-governed nation and each part had a role to play in its harmonious functioning.

Problems arose when the careful ordering of the body was disrupted, such as occurred when the stomach was overfed. Those who indulged too heartily, wrote the seventeenth-century physician Humphrey Brooke, risked 'making our Bellies sovereign to our Brains'.[78] In other words, to consume without manners and without restraint was to accept the illegitimate rule of the stomach over the steady governance of the head. The guts have been subject to suspicion since at least the Classical period of Ancient Greece, where the philosopher Plato split the body into two distinct regions, each associated with different emotional styles. While the head and heart were responsible for the 'higher' feelings – courage, honour, brotherliness – the lower organs, namely the liver, produced baser desires and appetites. Plato was not fond of the emotions, viewing them as a roiling and irrational force that had to be brought under control by the calm exercise of reason. To illustrate the dangers that such passions posed, the great philosopher looked to the stomach. Just as when the desire for food 'prevails over the higher reason and the other desires, it is called gluttony', so too are emotions felt and expressed without restraint to be condemned as grossly excessive.[79]

This muddying of bodily desires and emotional

expression shaped early Christian theology, under which certain feelings, such as envy, wrath and pride, were transformed into spiritual transgressions. At first glance 'gluttony' seems out of place among the other deadly sins – eating two portions of lunch hardly seems comparable to lusting after your neighbour's wife – but its presence alerts us to the way our emotional lives have long been imagined as bound, in complicated ways, with the appetites of the belly.[80] This is made clear by the emphasis that religious thinkers have placed on the need for discipline at the table; indeed, theological tracts on the sin of gluttony strongly resemble modern diet literature. Writing in eleventh-century Italy, the Dominican friar Thomas Aquinas warned that gluttony could take multiple forms. It was not enough to simply avoid excessive consumption, good Christians were also instructed to steer clear of foods that were too luxurious or daintily prepared, to eat only at designated mealtimes, slowly and without greedy eagerness.[81] Of course, where the latest diet fad might guarantee a flatter stomach or more energy, few would promise to secure the fate of your immortal soul. Early modern medicine, however, made explicit the connection between the everyday workings of the gut and the destiny of the transcendent spirit. Physical health helped to determine spiritual health to the degree that a disordered diet could compromise your relationship with God. As the writer William Vaughan enquired in the seventeenth century, 'Now for the soules faculties, how is it possible, but that the smoaky vapours which breathe from a fat and full paunch, should not interpose a dampish mist of dulnes betwixt the body and bodies light?' That Vaughan entreats his reader to take seriously the possibility that a fleshy belly might endanger the soul gives some insight into the long-standing relation of dietary restraint to moral purity. Concluding that 'all men which respect their bodies as the temple of the *Holy Ghost* [should] labour to keepe themselves pure, without repletion or surfet', he praises dietary moderation as a kind of devotion.[82] So intertwined were church pew and kitchen table that God might even be entreated upon to intervene directly in the process of digestion. In the sixteenth century the faithful were urged that because the Lord Almighty 'give strengthe to nature to digest', those who had fallen from His grace should expect food to be either 'filthily vomite[d] [. . .] up again' or else 'lie stinking in our bodies, as in a loathsome sinke or chanell, and so diversely infect the whole body'. Indigestion was, by this formulation, if not exactly divine punishment, certainly a sign that the eater had strayed from the path of righteousness.[83]

It is important to note, however, that the perfect Christian diet was not necessarily characterised by scarcity or privation. Instead, the virtues of temperance and moderation were praised, while the fasting practised by religious ascetics was generally discouraged as a marker of eccentricity. Someone like Roger Crab (Figure 7), the haberdasher turned hermit who became famous in the middle of the seventeenth century for eating only 'herbs and roots', was a subject of fascination, not emulation. This is clear from the publisher's note placed at the beginning of Crab's memoir *The English Hermite* (1655), which warns the reader not to try to follow his diet of 'poore homely food [. . .] as Corne, Bread, and bran, Hearbs,

Roger Crab *that feeds on Herbs and Roots is here;*
But I believe Diogenes had better Cheer.
Rara avis in terris.

Herbes and Roots

Deep things more I have to tell, but I shall now forbear,
Lest some in wrath against me swell & do my body tear.

Figure 7: Roger Crab, as printed in *Collection of Four Hundred Portraits of Remarkable, Eccentric and Notorious Personages* (1880)

Roots, Dock-leaves and Mallowes', and to instead strive for balance in all things.[84] Yet Crab, whose eventful life saw him fight for Oliver Cromwell in the English Civil War before being imprisoned and twice almost executed, was more concerned with the fate of his soul than the health of his body. In tune with a long-standing spiritual tradition that condemned the consumption of flesh as proof of humanity's corrupt nature and urged abstinence as the path to righteousness, he promoted vegetables as an antidote to man's sinful nature.[85] Following a strictly vegetable diet was also, he claimed, a way of cultivating mildness of character. While the meat of animals tended to produce aggression and other unmanageable passions in the eater, a natural bounty of 'Hearbs, Roots, Dock-leaves and Mallowes' calmed the temper and made it easier to lead a good Christian life. Despite his peripheral place in English society, Crab's linking of food to emotion was very much in line with hundreds of years of mainstream thinking on the role of diet in shaping individual temperament. From the medieval anatomists who used food to restore the balance of the body and the early Islamic philosopher Avicenna who observed that 'mental excitement or emotion' hinders digestion, from Jewish holy days like Yom Kippur in which fasting helps the devout to enter the right state of mind to atone, to the fifteenth-century monks who used fermented milk to quell anger and the private physicians who urged their patients to forgo rich meals in favour of even temperament, dietary advice often addressed itself to the emotions as well as to the stomach.

The relationship between the mind and the belly has been conceived of as a mutually informing one: good or bad digestion provoked certain feelings, but feelings could also shape the digestive process. This is not entirely divorced from some of the ways we think about the connection between food and mood today, ideas that have been shaped by relatively recent medical history. As Ian Miller, an expert in the history of the stomach, has summarised: 'Broadly speaking, nineteenth-century doctors tended to blame the gut for its effect on the mind. In the early twentieth century, psychologists were more likely to blame the mind for its effects on the guts', but either way they described the interconnection of brain and digestive organs in ways that resonate with how we conceive of that connection today.[86] Go back further, however, and a less familiar theory of how food works on the emotions starts to make itself known. Up until at least the eighteenth century, it was widely held that negative emotions were not only unpleasant to experience, but also frequently the cause of illness and even death. In Renaissance Europe, for instance, it was not uncommon for people to die from sudden shocks of joy and in seventeenth-century London 'lethargy', 'grief' and 'fright' appeared frequently in bills of mortality. Unruly passions could prove fatal, but it was possible to manage these through the workings of the gut as early modern doctors imparted food with the power to induce moods or even produce specific feelings – joy, anger, melancholy – in the eater. What this points to is a version of gut–brain communication more direct than is posited by modern science, one in which the imbibing of particular aliments produced preordained emotional states.

Eating was a dangerous business, an everyday act that welcomed foreign substances into the deep visceral interior, some imbued with potential to radically remake the mind and body. In 1584 Thomas Cogan, a Manchester physician and schoolmaster, published a guide to the perils of the dinner table titled *The Haven of Health*. Writing in English at a time in which most scientific literature appeared in Latin, Cogan wished to school his young students in the rudiments of everyday wellbeing, with chapters on 'Ayre, Meate and Drinke, Sleepe and watch, Labour and rest, emptiness and repletion, and Affections of the mind'.[87] Most space was given over, however, to a lengthy catalogue of common foods, their nature and their effects. Radishes, we are warned, cause belching; figs 'anoy the liver [and] fill the belly with winde'; and raw apples are best avoided as they cause indigestion, but can be safely consumed 'rosted, or baken, or stewed'.[88] Alongside physical consequences, he also details the impact that certain foods might have on the eater's state of mind. Beef can 'engendereth melancholy'; too much sugar inflames the temper; fried meat 'breedth choller' (irritability); ram's mutton should be left to those 'who would be rammish' (lustful); and turnips can, perhaps surprisingly, 'provoke carnall lust' in the eater.[89] Digestion could, by this account, transform character and remake something as seemingly fixed as personality, or as the polymathic jurist John Selden commented in his *Titles of Honor* (1614), the 'Mind's inclination follows the Body's Temperature'.[90]

At first glance *The Haven of Health* seems far removed from our current conception of how the interchange of stomach and mind operates. For Cogan, writing during the reign of Queen Elizabeth I, there was a direct causality to be drawn between consumption and emotion – eat beef and you will feel sad, eat the flesh of a ram and become horny like a ram. Today we would probably baulk at the simplicity of this idea, but it is widely accepted that diet impacts mood – the charity Mind recommends avoiding foods that increase blood sugar and the British Dietetic Association has outlined the ramifications of nutritional deficiency on mental health – and some of the advice offered in *The Haven of Health* chimes with contemporary wisdom. In a recent article for the *British Medical Journal*, an international group of scientists led by Joseph Firth of the University of Manchester reported the findings of their research into the relationship between mental well-being and daily diet. Among new insights into the role of the gut microbiome and the power of certain foods to boost immune activation, the researchers also point to the benefit of eating a well-rounded diet. Ideal is the so-called 'Mediterranean diet', which typically features a wide variety of fruits, vegetables and nuts and includes only the occasional consumption of red meat, and which is apparently associated with a reduced risk of depression.[91] Without becoming tangled in complexities here – perhaps those with better mental health feel able to eat better; how do medications used to treat psychiatric conditions alter appetite?; and where might class, gender or environment factor into this? – it is enough to note that what is consistently pressed upon us by healthcare professionals, diet gurus and cutting-edge science is the need to achieve some kind of healthy balance. Not only are we encouraged to eat a varied diet, but we are also warned that white sugar is likely to cause a spike in energy followed by a steep decline, that too much cheese raises blood cholesterol levels as the liver struggles to process fat and even that indulging in a glut of fresh fruit overburdens the digestive system, resulting in bloating, if not a swift trip to the loo. What bridges the gap between the sixteenth-century *Haven of Health* and modern nutritional research, then, is a common investment in the ideal of bodily equilibrium.

The association of health with balance can be traced back to a system of medicine that originated in the ancient world, and which only fell out of favour towards the nineteenth century. Humorism dictated that the body was governed by four vital fluids – blood, yellow bile, phlegm and black bile – that were distinct in character – hot, dry, wet and cold – and which corresponded to the four elements – air, fire, water and earth. These four fluids also produced four temperaments: blood was associated with sanguinary (cheerfulness), yellow bile with choleric (irritability), phlegm with phlegmatic (reserve) and black bile with melancholic (depression). The key to good health was to keep the humours in balance by adjusting environmental factors and modifying diet. Ideally a person should be a little warm and a little moist, so dietary correctives were usually aimed at moving the eater closer to this perfect mix. The elderly were thought to be prone to dryness, so those of a certain age would be encouraged to consume onions and squashes, both thought to be classically 'wet' foods, while young men who sometimes ran too hot would be entreated to eat 'cold' foods like lettuce or fish. Digestion was, according to this ancient theory, one of the most important processes in the body, as it involved not only the assimilation of food but also the production of spirits and humours. The stomach and digestion were linked to mental and emotional experience since the proportion of these humours within an individual affected temperament, making a person melancholic, choleric, phlegmatic or sanguine accordingly.

The interdependence of diet and environment in shaping emotional wellbeing is clearly illustrated by Cogan's chapter 'Of Biefe', in which he begins by acknowledging Galen as the founder of humoral medicine before taking issue with his characterisation of beef as a melancholic meat. As a meat that was cold and dry it was not usually recommended for those already prone to an excess of black bile, and along with low mood beef was thought to cause unsightly skin conditions like 'cankers, scabbes, leprie'. Though Cogan does not take issue with the notion that the meat can sometimes 'maketh grosse bloud and engendereth melancholy', he questions whether advice formulated in the heat of southern Europe should be directly applied to the northern climes. The cold weather of England, Cogan wrote, 'doth fortifie digestion' so that 'stronger nourishment' is required by its residents and as such beef ought to be recommended to those that 'be lustie and strong'.[92] Humoral balance, then, could only be achieved by considering factors as various as the weather, the quality of the meat, the age of the animal and individual predisposition.

Complicating the situation further was the belief that the stomach itself possessed a peculiar humoral makeup. The organ was thought to be dominated by the hot, dry yellow bile associated with choler. And as such it was viewed as the site of irascible passions like pride, stubbornness and anger. The analogy between the gut and bad temper was sustained in popular language, as historian Jan Purnis outlines: 'The verb form of *stomach* and many now-obsolete forms including *stomachate, stomachous, stomachful* and *stomaching* were most often connected to feelings of anger and resentment.'[93] So established was the connection between disordered emotions and pathological digestion that medical thinkers began to publish books on dining etiquette. In the early seventeenth century a well-known French physician named Joseph Duchesne wrote a treatise on the subject. It was imperative, he urged, never to come to the table upset because anger can singe food as it passes through the body and fear can halt the digestive process altogether. We must, therefore, control our emotions at dinner, not only in the interest of manners and civility, but also to ensure the continued health of the body.[94]

This idea had a long life. Writing in 1872, an American health reformer named Diocletian Lewis insisted that:

Our manners at the table have much to do with our digestion. Politeness must be set down among the means of a healthy stomach. In the first place, if we offer the bread, the butter and the sauce to others, we interrupt the otherwise unbroken shovelling-in business, and thus make eating more deliberate, which is an advantage; and in the second place, the temper induced by this mutual kindness is eminently favourable to the stomach functions. A kind action always tends toward health; an unkind or mean one tends in the opposite direction. This is a general law, and especially applicable to table manners.[95]

Born into a poor farming family in 1823 and working in a cotton factory at the tender age of twelve, Lewis educated himself and went on to have a remarkable career as a prison doctor, celebrity homeopath and physical culturalist. Later in life, he took up the cause of temperance and joined the crusade against alcohol that would eventually lead to Prohibition in 1920. One tactic that he favoured involved gathering groups of church-going women to pray and sing hymns outside bars with the aim of shaming patrons into giving up the demon drink. This proved so successful that – according to Lewis – within three months more than two hundred towns in Ohio were liquor-free. Over the next few years the crusade made inroads across the country. The trouble with alcohol consumption, for Lewis, was not only that it caused all sorts of terrible health problems, but also that it impaired the moral sense of the drinker. One could not, he insisted, lead a decent Christian life and resist worldly temptations while under the influence of dangerous intoxicants. Despite being directed at digestion rather than drink, Lewis's thoughts on dining etiquette speak to a similar set of anxieties. Consuming the right foods in the right way was, for him, a method for taming the passions and quelling baser desires. Instead of giving into the insistent demands of the hungry belly and 'shovelling-in' the meal in front of them, the diner should demonstrate restraint and self-control.

For Lewis, table manners disciplined the eater and ensured that the demands of the noisy stomach remained subservient to the will of the rational mind. Like most of the ideas and individuals featured in this chapter, he conceived of the gut and its appetites as forces that needed to be kept under control. By following a dietary regimen or adhering to a strict code of etiquette, 'ugliness' at the table might be avoided and order maintained. Rules are, of course, there to be broken. Think of the polite dinner that descends into a wild kitchen dance party or the guest who feels compelled to share – loudly – the intimate details of their sex life after one glass of wine; consider the eviscerating arguments that erupt over candle-lit anniversary celebrations or the messy, sticky chaos of sharing a meal with anyone under the age of five. Etiquette may reflect an ideal rather than a lived reality, but it still wields great power as a prescription for how to behave in company. Containing the disruptive corporeality of slurping, burping and farting, manners help to establish the social order by carefully regulating the body. Flaunting the laws of the table, then, can perhaps be read as an act of sedition. In a well-known study of the satirist François Rabelais, the Russian philosopher Mikhail Bakhtin identified the Renaissance carnival as an event that upended hierarchies of all kinds. He observed that 'as opposed to the official feast, one might say that the carnival celebrated temporary liberation from the prevailing truth and the established order; it marked the suspension of all hierarchical rank, privileges, norms and prohibitions'.[96] The carnival was a celebration of what Bakhtin describes as the 'grotesque body', one that was defined by eating, drinking, sex and pregnancy, rather than by so-called 'higher functions' like reasoning. Wriggling free of the limits and strictures set by popular convention, this 'grotesque body' revelled in the messiness of consumption, digestion and excretion. That it was only permitted to do so at certain times of the

year, however, reveals something of how dangerous breaking the rules of the table can be.

This chapter has explored how we have tried to control
the gut, to manage its influence on the rest of the body
and on mental wellbeing. Expanding on this examination
of the historical linking of mind and belly, the following
chapters look more closely at the process of digestion and
consider why this appears to have been so often linked to
understandings of labour and work. The complexities of
consumption will come up again in *Rumbles*, but here it
is perhaps enough to note what remarkable powers have
been ascribed to a change of diet: it can improve health,
it can rebalance the body, it can produce certain feelings
or states of mind and it can even transform one's relationship to God. All the attention that has been directed at
the gut – how to control it, manage it, tame it, civilise it –
only serves to underline how powerful the organ is, what
a remarkable influence it seems to exercise in the world.
Nowhere is this clearer than in relation to the emotions
and the last three chapters have revealed a long-standing
entanglement between feeling and digesting. Variously
imagined as biological, physiological, religious, psychological, cultural and social, this relationship resists simple
classification, but it does point to the remarkable prominence of the gut in shaping understandings of the self
through history.

WORK

Throughout the middle years of the Weimer Republic, a period of intense industrial development, rapid
cultural change and fizzing intellectual ferment, the Ger-
man-speaking world was enthralled by a popular series
of books on the human body. Published in five volumes
between 1922 and 1931, *Das Leben des Menschen* or *The
Life of Man* was the work of Fritz Kahn, a German-Jewish
physician who was committed to making the wonders of
science and medicine available to the broadest audience
possible. Part of the appeal of his books lay with their use
of illustrations, visually arresting and intricately realised,
to convey complex information about our inner workings.
In the best-known of these, *Der Mensch als Industriepalast*
or *Man as Industrial Palace* (1926), the body is depicted
as a well-ordered factory: in the brain, workers stand at
panels that control the central nervous and respiratory systems; the spinal cord is a telephone line presided over by
diligent switchboard operators; in the lungs oxygen and
carbon dioxide travel along gleaming automated pipes;
the thyroid gland is a large industrial silo and the heart
an engine room of pumping pistons (Figure 8). By far the
largest departments in this 'industrial palace' are those

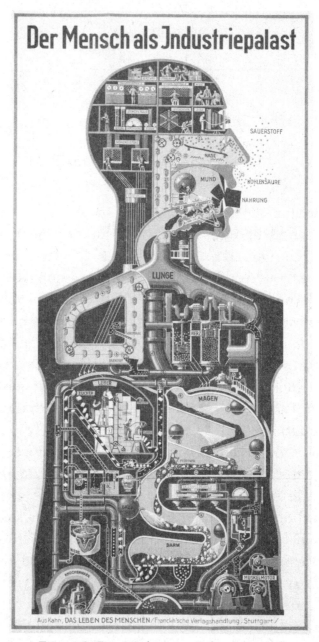

Figure 8: Fritz Kahn's 1926 depiction of
the body as an 'industrial palace'

devoted to the digestive process. Beginning in the mouth where a hand-operated saliva pump eases the flow of food from the teeth towards a yellow tube labelled *Speiseröhre*, literally 'food pipe', the German word for oesophagus, to the conveyor-belt stomach where the food is to be treated by sprinklers dispensing hydrochloric acid, pepsin and ferments, before being discharged into the intestine to be broken down by bile pumped from the liver and a cocktail of juices from the pancreas, then finally expelled through the colon.

While the human factory is very much a product of the early twentieth century, a futuristic vision of sleek modern efficiency rendered in the visual language of the Bauhaus, the idea of the body as a site of daily work has a longer history. This is particularly true of digestion, a reflection – at least in part – of how labour intensive the process of consumption, absorption and defecation is. Beginning in the mouth with the crushing and grinding of the teeth, then down through the gastrointestinal tract where food mixes with digestive juices that dissolve it into smaller molecules able to pass through the walls of the small intestine into the bloodstream, then finally on to the large intestine and out of the body. It is a procedure that involves millions of enzymes, several different kinds of acid, the precise action of dedicated muscles, the coordination of multiple organs and the careful oversight of the parasympathetic nervous system. More so than other bodily process, digestion can feel like work: consider the ill-advised late-night kebab that keeps you up all night or the hearty bean salad that your gut seems to grapple with for days. The following chapters examine what the connection between work and digestion might mean for its broader cultural history. Beginning with the puzzle of how the gut works, before examining its relationship to the work of the brain and closing with the history of the working lunch, they explore what it means to have a stomach that labours.

4

How Does the Gut Work?

On 6 June 1822 an accident took place on Mackinac Island. Located in Lake Huron, the vast body of water that borders the US state of Michigan and the Canadian province of Ontario, Mackinac Island had served as a fur trading post since the early seventeenth century, one that was uniquely well positioned between the frozen north and the growing commodity markets of the south. By the beginning of the nineteenth century the island was in the hands of the American Fur Company, and it was in one of their supply stores, as customers jostled among the buckskin coats and pairs of moccasins, that the calamity occurred. Gurdon Saltonstall Hubbard, an influential Chicago land speculator, was witness to the incident and later recalled the scene:

> This St. Martin was at the time one of the American Fur Company's engages, who, with quite a number of others, was in the store. One of the party was holding a shotgun, which was accidentally discharged, the whole charge entering St. Martin's body. The muzzle was not over three feet from him – I think not over two. The wadding entered, as well as pieces of his clothing; his shirt took fire; he fell, as we supposed, dead.[97]

The unlucky shopper, Alexis St. Martin, was a young French-Canadian *voyageur*, a canoe operator, sent into town to deliver beaver pelts. As he lay unconscious on the floor of the shop someone sent for the only doctor in town, a surgeon by the name of William Beaumont who was stationed at a nearby military outpost. On initial examination it seemed unlikely that St. Martin would recover from his injuries, but after a few days he seemed to rally, and Beaumont was able to undertake a more detailed examination of the wound. The bullet, he discovered, had ripped through the patient's abdomen, fracturing ribs, lacerating muscles and puncturing a lung as it went. What was remarkable, as he recorded in his case notes, was that through the hole it was possible to glimpse the 'stomach with a puncture in the protruding portion large enough to receive my forefinger, and through which a portion of his food that he had taken for breakfast had come out and lodged among his apparel'.[98] For several weeks, anything consumed by St. Martin re-emerged in half-masticated form through this opening and even when he was able to digest again, the gastric fistula that had formed between the wound and the stomach remained open.

This fleshy window presented Beaumont with a unique opportunity to witness the digestive process at work. Carefully attaching small pieces of food to lengths of string, he would lower the morsels through the hole and directly into the stomach, removing them at intervals to monitor their decomposition and take acid samples. Similar experiments had been undertaken with live animals, most famously by the eighteenth-century Italian biologist Lazzaro Spallanzani, who set out to prove that digestion involved some kind of chemical operation using his pet falcons as test subjects. Suspending bits of meat through the upper digestive tract and into the stomachs of the unfortunate birds, he was able to measure the rate at which the food was broken down. In doing so, he was attempting to upend the medical orthodoxy of his day, which insisted that digestion was a mechanical process that involved the steady grinding down of food. Spallanzani wanted to prove that the acids that swirled around in the stomach were far more significant than previously acknowledged. While his close contemporary the French entomologist René Antoine Ferchault de Réaumur claimed that the 'gastric juice has no more effect out of the living body in dissolving or digesting food than water, mucilage, milk, or any other bland fluid', Spallanzani determined that in fact these juices – and not chomping teeth or roiling guts – were responsible for breaking down food in the body.[99] By the time of St. Martin's unfortunate accident, it was generally accepted that gastric acid was essential to digestion, but its actions inside the body remained a mystery. What made the unfortunate trapper's fistula so irresistible was that it provided, for maybe the first time, a live demonstration of the human stomach in action.

Today it is the brain that baffles and bewitches the medical imagination, but through most of history the stomach was held to be the most enigmatic of organs. Obscured by the liver, nestled by the gallbladder, spleen, pancreas and large intestine, it hides from prying eyes and pressing fingers. More testing than its remote location is the fact that the organ only really makes sense in motion. While the post-mortem dissection of the heart, for instance, reveals the intricate structure of valves, chambers and ventricles, removed from the context of the living body the stomach's functions are less apparent. From the eighteenth century, as dissection came to occupy an increasingly central position in medical education, instructors were forced to find creative ways to teach with this most unobliging of organs. Edward Jenner, now best remembered for pioneering the concept of vaccination, was also a talented anatomist, as this elegantly dissected organ from the late eighteenth century attests to (Figure 9). Resembling a gruesome hand fan, this specimen is a section of stomach flattened and injected with wax to show the veins, arteries and delicate membrane of its wall. Though an invaluable aid to better understanding its nervous structure, Jenner's preparation reveals little of the purpose of the organ, how it operates in the living or behaves in relation to the rest of the digestive system. It was not until the invention of technologies like the X-ray and the gastrograph at the turn of the twentieth century that the work of the stomach could be fully observed and systematically measured.

This lag in medical knowledge allowed the rich metaphorical language of the gut to flourish, a digestive imaginary that shifted across time and space. Writing in 1884, Alexander Blyth, the Medical Officer of Health for Marylebone, described the role of the stomach in the 'city of the body'. In this 'peripatetic city' there are 'highways' along which 'oxygen-laden boats' travel and each citizen has his strictly 'appointed place and vocation', from the workers in the liver busy manufacturing 'bile and glycogen', to the 'officer' who sits below a 'thatch of hair'

Figure 9: Part of a human stomach dissected by Edward Jenner

carefully directing the intelligence. This is a bustling metropolis that 'never rests' and the complex 'internal work' of the body relies upon the speedy 'distribution' of food among its many inhabitants, as different jobs in the 'city of the human *Bios*' demand different forms of nourishment.[100] Given that Blyth wrote from the perspective of public health and in a period of unprecedented urbanisation, it is unsurprising that he would look to the well-ordered metropolis as a fitting image for the body. By contrast, the early modern stomach was more often described as a cauldron where the meals needed to fuel the rest of the body were cooked. As William Vaughan remarked in his popular medical treatise *Directions for Health* (1600), 'Our stomake is our bodies kitchin.'[101] Elsewhere, one of the foundational texts of modern human anatomy, Andreas Vesalius's *On the Fabric of the Human Body* (1543), described digestion as a process akin to preparing a meal. Once chewed and swallowed food 'passeth to the next Guts as to a gate' where the 'heat of the Bowels round the Stomach', aided by the fire of the liver, cooks it like a stew in a pot and readies it for distribution around the body.[102] Likewise, in Edmund Spenser's *The Faerie Queene* (1590) the stomach is depicted as 'the Cauldron many Cookes accoyld/ With hookes and ladles, as need did require', while 'Digestion' is the 'Clerke' charged with managing the 'kitchin'.[103] The stomach, then, was not only envisioned as a passive receptable for half-chewed vitals, but also sometimes likened to the busy kitchen of a great house, labouring tirelessly to keep its occupants fed.

In the early sixteenth century, the German-Swiss alchemist and mystic Paracelsus proposed that the stomach should be understood not as a kitchen, but rather as a chemical laboratory. In the *Book of Tartaric Diseases* (1537) he claimed, against the dominant humoral model of his day, that digestion was caused by acid acting on food in the stomach. A century later, another alchemist, the Flemish physician Jan Baptiste van Helmont, returned to this idea and offered the first comprehensive chemical account of digestion. An iatrochemist – a branch of sixteenth-century medicine, grounded in the principles of alchemy, but which also drew on recent discoveries in chemistry and biology – van Helmont characterised digestion as a kind of fermentation, taking place in the stomach but also involving the gallbladder and the spleen. Foreshadowing Lazzaro Spallanzani's experiments on falcons, van Helmont's interest was apparently first sparked when, as a young boy, a tame sparrow who often sat on his shoulder attempted to bite his tongue. Having noticed how tart the mouth of the unusually bold bird was, he became convinced that acid must play some role in the breakdown of food and undertook an experiment to prove his hypothesis. Settling a small piece of meat first in a solution of vinegar, next of lemon juice and then of leavened flour, he waited for the acids to do their work, and was sorely disappointed when the morsel remained intact. Undeterred, he proposed that the body itself must contain a multitude of chemical substances responsible for breaking down different types of food, species specific and particular to the individual. In other words, he came close to discovering the enzymes, the lipases that gather fatty acids, proteases that split proteins into peptides and amylases responsible for carbohydrates, that are now so essential to our understanding of the digestive process. Less familiar to modern science, however, is the alchemical principle that governed his lengthy investigations. Following Paracelsus, he ascribed to the idea of *archaeus*, a kind of vital force that governed all living things and whose chief seat in the body lay deep in the stomach. According to this principle, digestion was both physical and metaphysical, a process that involved material and spiritual forces. To understand the mechanics of that process was, for alchemists like van Helmont, to begin to unravel the mysteries of the universe.

This was not the only time that the secrets of life were sought in the gut. At the turn of the nineteenth century the stomach was called upon as evidence in a debate over the nature of life itself that was raging through the British scientific establishment. In 1816 William Lawrence, recently elected Professor of Anatomy and Surgery, was commissioned to deliver a series of lectures to mark the opening of a new museum at the Royal College of Surgeons in London. The youthful Lawrence recognised this as a unique opportunity to make his mark on the field and, courting controversy, he embarked on a lengthy discourse on the nature of consciousness in which he asserted that mental processes were a function not of the soul or ineffable spirit, but of the brain. Scandalised by its implications – that man was not divine creation but crudely animated flesh – Lawrence's research was widely condemned. Among his most voracious critics was John Abernethy, a surgeon at St Bartholomew's Hospital who had previously engaged Lawrence as a demonstrator on his own anatomy course. Offended by the 'materialism' espoused by his former employee, Abernethy encouraged the 'thinking people' of England not to be swayed by new ideas about the mind that 'subvert morality, benevolence, and the social interests of mankind'.[104] The argument between the two former colleagues eventually breached the walls of the Royal College of Surgeons, enflaming public debate over the nature of life, religious truth and the proper scope of medical investigation.

According to the literary scholar Marilyn Butler, the controversy also inspired Mary Shelley's famous depiction of overweening scientific ambition in *Frankenstein; or, The Modern Prometheus* (1818).[105] The genesis of the book, famously composed during a round of ghost stories told in a house rented by Lord Byron on Lake Geneva, coincided with the dispute; as Butler writes, 'there surely cannot be much doubt that the group were speaking of the [. . .] debate between Lawrence and Abernethy'.[106] The creature, an assemblage of dead human and animal parts, reanimated by a combination of chemicals and electricity, raised the same questions as the quarrelsome surgeons: what makes us human and where does life come from? For Lawrence, who was also Percy Shelley's personal physician and well acquainted with the social set that gathered at Lake Geneva that summer, the answer lay in the structure of the body itself, which like any other machine depends upon the cooperation of all of its parts. In contrast, Abernethy argued that life was not only anatomical arrangement, but that it existed as a kind of vital principle, a 'mobile, invisible substance' external to the body. This vital spark was essentially the Christian soul rendered in the language of modern science and so to deny its existence was to refuse the promise of eternal life.

The stakes, in other words, could not be higher. For Lawrence, Abernethy was like the 'poor Indian, whose untutored mind/ Sees God in the clouds, and hears him in the wind'; while for Abernethy, Lawrence was a dangerous atheist determined to reduce the human body to a mass of twitching flesh.[107] Interestingly, though the two clearly held very different views of the world, they were both trained by the same man and both believed themselves to be faithfully following his teachings. Their progenitor was the Scottish surgeon John Hunter, who had throughout the second half of the eighteenth century distinguished himself as a skilled surgeon and a pugnacious advocate of the scientific method. Having moved to London to train under his elder brother, the renowned obstetrician William Hunter, he was elected Surgeon to St George's

Hospital in 1768, built up a thriving private practice and established a vast teaching collection of over 14,000 preparations, which included more than 500 different species of plants and animals, alongside human specimens. In his lectures, Hunter formulated a complex set of ideas regarding the interplay between the structure and functions of the body. After his death, his former pupils squabbled over the implications of his theories, with Lawrence insisting that they implied that life involved only a particular organisation of material and Abernethy countering that in fact they suggested that life could not be understood in terms of the 'matter of animals and vegetables' alone.[108]

When called upon to substantiate this claim and provide evidence of the action of an animating power of some kind, it is significant that Abernethy looked not to the brain but to the belly. As George Cruikshank satirised in his 1828 cartoon (Figure 10), Abernethy was known to his patients and the wider public for his obsession with the vagaries of the digestive process. Believing that poor digestion lay at the heart of many common conditions, he regularly prescribed dietary change as the surest route to good health. However, his interest in the gut extended far beyond the question of what to eat, because for Abernethy the stomach held the key to understanding the nature of life itself. This conviction originated in the teachings of his mentor. Dissecting the stomachs of dead animals, Hunter had noted that bits of food lingered on undigested and elsewhere he conducted experiments that demonstrated that acids in the stomach did not burn the delicate living tissue that surrounded them. Though he himself remained convinced that digestion was primarily a chemical process,

Figure 10: John Abernethy's obsession with the
gut satirised by George Cruikshank, 1828

concluding that 'it is neither a mechanical power, nor contractions of the stomach, nor heat, but something secreted
in the coats of the stomach, and thrown into its cavity,
which there animalizes the food', Hunter's research also
seemed to suggest that animated matter possessed the
power to suspend or initiate that process.[109] The tantalising possibility that this opened up, for Abernethy and
other like-minded scientists, was that the work of the
stomach was set in motion by some exterior spark of life.

Over the course of his long career, Hunter dissected
hundreds of stomachs and gained useful insights into
the mystery of the organ's operations. To carry out this
work and to teach the next generation of surgeons he
required a fairly steady supply of fresh cadavers. Only by

direct observation and dissection could progress in the field of medicine be made, as his elder brother William insisted: 'It is by Anatomy alone, that we know the true nature, and therefore the most proper cure of the greatest number of local diseases.'[110] Up until the eighteenth century, anatomy in Britain had relied upon royal decree: in 1506 King James IV of Scotland granted the Edinburgh Guild of Surgeons the right to dissect a number of executed criminals, Henry VIII donated the corpses of four hanged men to the Company of Barber-Surgeons and over the years many similar 'gifts' were made to scientific organisations like the Royal Society. With the founding of new medical schools in Edinburgh and London, however, demand began to outstrip supply. The so-called 'Murder Act' of 1752 attempted to address this shortage by authorising judges to extend the convicted criminal's sentence beyond death, from the gibbet to the anatomy theatre. Still the need for corpses could not be met within the confines of the law and body snatchers were hired to steal bodies from their graves. Several 'resurrectionists', as they came to be known, streamlined the process by murdering their chosen cadaver and then delivering it fresh to the surgeon.[111]

The Hunter brothers were key figures in this gruesome episode in the history of medicine. In an engraving from 1773 (Figure 11) depicting the capture of a body snatcher by a nightwatchman, as a female corpse in its burial shroud spills out from a hamper, William Hunter is identified as the fleeing anatomist by a dropped piece of paper on which it is possible to make out the fragment 'Hunter's lectu—'. Tasked with stocking the dissecting tables of his

Figure 11: This 1773 engraving by W. Austin implicates
William Hunter in the crime of body snatching

private anatomy school, he was known to have engaged
the services of grave robbers, while it was rumoured
that his younger brother preferred to do the dirty work
himself. John Hunter is today perhaps best remembered
for the ruthlessness with which he pursued Charles Byrne,
an exceptionally tall man – seven foot six inches to be
precise – known to the public as the 'Irish Giant'. Soon
after his arrival in London, Hunter approached him and
offered to pay for his corpse so that he could display it as
an unusual specimen in his private museum. Understand-
ably horrified by this gruesome proposition and fearing
that the anatomist would not take no for an answer, Byrne
arranged that on his death his body would be sealed in a

lead coffin and buried at sea. He was, it transpired, right to be paranoid, as Hunter arranged for the cadaver to be snatched before it reached the ship and put his skeleton on display at the Hunterian Museum in London, where it remained for two hundred years before recently being removed from public view.

Our knowledge of how the gut works is bound up with this complex and compromising history. The techniques and methods developed on the dissecting table opened the stomach up to sustained scientific observation, but in the process the individual was often dehumanised and transformed into a scientific object. Nowhere was this clearer than in the case of the unfortunate Alexis St. Martin who we encountered at the beginning of this chapter. Having survived being shot at close range, he found himself without employment and too unwell to make the journey back to his family in Quebec. Sensing an opportunity, William Beaumont offered to employ St. Martin as a servant in his home on the condition that he submit to further examination. The arrangement was ratified by a contract that permitted the doctor to undertake 'Physiological or Medical experiments' and required his patient to acquiesce in 'the exhibiting and showing of his said Stomach'.[112] Being illiterate, St. Martin was likely unaware of the document's implications and could not have foreseen the strange course it would set him on. Over the next ten years, Beaumont – as one of his many correspondents remarked – transformed the French-Canadian into a 'human test tube' with which to probe the mysteries of digestion.[113] Using a length of elastic tubing to extract gastric juice directly from the stomach, gathering vials of fluid to measure the decomposition of different foodstuffs and at one point licking the walls of the stomach to judge its acidity, he recognised his servant to be a valuable experimental subject, one that he was apparently reluctant to give up. Not only did the doctor refuse to do what he had promised in the first place and close the gastric fistula, when his patient finally tired of the endless probing and absconded, Beaumont hounded him with offers of more money and threats of dire legal consequences. Such was his persistence, that when St. Martin died aged seventy-eight in 1880 his family waited for his corpse to decompose before burial, to deter body snatchers from digging it up to sell to the anatomists. Given that Beaumont's fevered pursuit was driven by the possibility of exhibiting his reluctant patient as part of a lecture tour through Europe and that after his death another physician named William Osler tried to secure his stomach for the Army Medical Museum in Washington, St. Martin's concerns seem to have been – like those of the so-called 'Irish Giant' – more than justified.

In 1833 Beaumont's *Experiments and Observations on the Gastric Juice, and the Physiology of Digestion* was published – the result of a decade's worth of ethically dubious experimentation. Alongside a detailed record of the two hundred plus experiments he had conducted, Beaumont set out his thinking on the physiology of the gastric system, touching on topics as diverse as the effect of wine on the stomach's mucus membrane, the necessity of condiments and the digestibility of fish, to the impact of hunger and satiety on the appearance of the stomach. Most prominent, however, is the subject of gastric juice.

Following in the footsteps of pioneering enteric explorers like Paracelsus and Jan Baptiste van Helmont, Beaumont proved that digestion was a chemical process and set out to understand the composition and action of stomach acid. *Experiments and Observations* broke new ground by demonstrating the impact of temperature, distinguishing gastric juice from mucus and providing evidence for the chemist William Prout's previous discovery that hydrochloric acid was likely one of its essential components. Beaumont also observed that mental disturbance had a detrimental effect on digestion. 'Strong mental emotions, such as anger', were, he wrote, likely to diminish the 'secretion of gastric juice' and thus impair the gut's usual operations.[114] Exactly what the surgeon did to provoke anger in his patient is not, however, disclosed to the reader. Today Beaumont is hailed as the 'father of gastric physiology' for his pioneering research into the mysteries of the stomach, but *Experiments and Observations* was not a commercial success. As writer Mary Roach has uncovered from her research with the William Beaumont Papers at Becker Medical Library, it was such a flop that its author was reduced to hustling friends into buying copies and pressing reluctant publishers to bring out a British edition.[115] Finally, the Scottish firm Maclachlan and Stewart picked the book up and put out a new edition in 1838, accompanied by an introduction from a prominent Edinburgh physician, Andrew Combe. Even here, however, Beaumont's achievements were overshadowed by the celebrity of his patient and Combe closed his preface by entreating the Royal Society to use 'their influence and means to have St Martin brought over to this country' so that a fuller inspection of the remarkable fistula might be undertaken by a more qualified 'committee'.[116] As Beaumont eventually came to realise, his in-depth scientific recitations on the exact chemical composition of gastric fluid were of little interest without the accompanying spectacle of a man with a gaping hole in his belly.

Though his research did not receive the recognition he thought it deserved, Beaumont's experiments on St. Martin do have an important place in the cultural history of the gut. In 1933, to mark the centenary of the publication of *Experiments and Observations*, the Beaumont Foundation held a series of commemorative lectures. One of these was delivered by the Harvard physiologist Walter B. Cannon, who opened by disclosing a personal connection: he was born in 1871 in Prairie du Chien in Wisconsin, the settlement in which Beaumont had conducted his famous experiments. These were, he continued, made all the more remarkable given that they were conducted in a 'backwoods army post, without laboratory aids, with no journals or literature to consult, with no associates to encourage him' and involved carefully cajoling the 'fistulous Alexis' into participating. Happily, Cannon concluded, one no longer had to be a rugged frontiersman to conduct research into the gut, as a wonderful new technology had afforded 'every physician' the opportunity to be a 'William Beaumont and his patient an Alexis St. Martin'.[117] In the late nineteenth century, while he was still in medical school, Cannon used an X-ray machine to study digestion in animals (Figure 12). By utilising a form of electromagnetic radiation known as Röntgen rays, named after the German physicist who discovered them in 1895, he was able to observe gastrointestinal movements without a serendipitous gunshot wound.

He began his investigations with a goose, fed with food mixed with bismuth – a chemical element opaque to X-rays – and shut in a box to keep its neck straight giving it 'a most absurd and pompous air'. Using the machine, Cannon was able to chart the movement of the food through the oesophagus, stomach and intestines of the goose, measuring peristalsis as it worked its way through the system. Encouraged, the young physiologist broadened the scope of his experiments to include male and female cats. Differences between the sexes quickly became apparent: under restraint male cats ceased the digestive process, while in females the 'peristaltic waves took their normal course'. Looking to explain this discrepancy, Cannon initially attributed the cessation of peristalsis in the male cats to the violence with which they resisted the X-ray machine, but later came to understand it as being a result of emotional stress. Having found that frightened cats produced less gastric juice, but that reassuringly stroking the fur of a distressed cat helped to restore gastric mobility, he concluded that 'stomach movements are inhibited whenever the cat shows signs of anxiety, rage, or distress'.[118] From these early studies in digestion, Cannon emerged as a leading expert in the sympathetic nervous system, whose work uncovered the connections between the physiological processes of the body and different emotional states. In books like *Bodily Changes in Pain, Hunger, Fear and Rage* (1915) and *The Wisdom of the Body* (1932), he established himself as a leading proponent of psychosomatic medicine.

It is significant that the organ credited with having

Figure 12: Photograph of the X-ray machine used by
Walter Cannon in his investigations into swallowing

convinced Cannon of the inseparability of mind and body was the stomach. Such acclaim speaks, in part, to the organ's increased visibility at the time he was working. The X-ray was one of several late nineteenth-century technologies, including the stomach bucket, gastrograph and the gastro-investigative tube, that made the activities of the guts newly legible to scientists. As Cannon enthused in his lecture to the Beaumont Foundation, 'it is now possible by means of the Röntgen rays to examine readily in man the activities of the gastric wall, the changing contour as disease may influence it, and the effects on digestion of many factors that were first observed in lower animals.'[119] Yet even as its actions became ever more visible, as new techniques allowed scientists to witness the constriction and relaxation of muscles as food moves slowly down the digestive tract, the exact nature of the gut's relation to the mind remained unclear. Our current fascination with the workings of the enteric nervous system and the intricate operations of the vagus nerve is perhaps a response to this gap in medical knowledge, but it is also – perhaps more importantly – a reflection of the concerns and preoccupations of our peculiar historical context. In this we are not alone. Cannon's investigations into the autonomic nervous system, which would eventually lead him to coin the term 'fight or flight' to describe our response to a perceived threat, were conducted as the wider scientific world turned its attention to the question of stress and disease.

In the early twentieth century Cannon examined the stomach to better understand the impact of stress on the body, while sixteenth-century alchemists looked to the gut to unravel the secrets of the universe, and nineteenth-century anatomists searched the entrails for signs of God's creative power. What unites these enquiries is a shared conviction that by unmasking the mysteries of digestion and exposing the most enigmatic of organs we might gain insight into larger truths about the mind, the body and even the nature of human existence. Along the way physicians like Beaumont and Hunter overstepped the boundaries of decency, denying the humanity of their research 'subjects' in the name of scientific knowledge. Such overweening ambitions have often been frustrated by the work-a-day nature of the organ itself. The gut is devoted to the steady processes of ingestion, breakdown, absorption and elimination; it is not so easily romanticised as the heart. This may explain, in a roundabout way, why we have so long been convinced of some special connection between the gut and the brain. Both perform a kind of work in the body that feels perceptible: the grind of the stomach as it tries to digest a too large meal and the strain of the mind as it wrestles to comprehend a difficult concept. Moving on from the idea of 'discovering' the gut, the next chapter returns to the vexed question of its relationship to the brain and considers why this connection has so often been imagined in terms of their shared work ethic. What does it mean to be creatively constipated? Can you think and digest at the same time? And why might it not be wise to read at the dinner table?

5

Brain Work

The secret to good health lies not in what you eat, but in how you eat it. According to a study from psychologists based at Birmingham City University, obesity is the product of chronic inattention.[120] Distracted by phones, social media and television, we have, the researchers argue, become increasingly deaf to our stomach's signals and unwilling to listen when it is telling us to stop eating. The solution, apparently, lies in adopting more 'mindful' eating practices. These include chewing each mouthful of food slowly, pausing in between bites, taking time to digest and, most importantly, ensuring that you finish your meal uninterrupted by online shopping or the latest Twitter spat. Despite being couched in the language of contemporary science – disrupted neural transmitters and the parasympathetic nervous system – this research concerns itself with a centuries-old problem: what if the brain and the digestive system were not only in communication with each other, but in direct competition? In the injunction to practise mindfulness at the dinner table, to focus solely on the acts of tasting, chewing and swallowing, lies the tacit assumption that activities that engage the intellect somehow impair the work of the body. Today we think about this problem primarily in terms of attention, with poor eating habits attributed to the modern information overload, but history offers up other models for understanding the relationship between the work of the gut and the work of the brain. There are, for one, different ways to think about the question of attention itself. Creativity, for instance, could be thought of as a product of inattention, really, of allowing the mind to wander hither and thither, unimpeded by the demands of concentration. Good ideas often come up over lunch, whether in conversation with friends or simply staring off into the middle distance between bites of a sandwich. And who can deny the worthwhileness of ice cream and daydreaming on a hot summer's day?

Building on earlier discussions of the gut–brain axis, this chapter devotes itself to better understanding the connection between digesting and thinking. The two can often be found in semantic entanglement with one another: think of common phrases like 'food for thought' and 'hungry for information' or the way that bad news might be 'hard to digest'. In Spanish the word *tripas* can mean both 'intestines' and the substance of a text, much like the use of 'meat' in English, and in French the expression *donner du grain à moudre*, to 'give grain to the mill', means to chew over an idea. Tracking meaning back and forth between the body and the social world, these expressions point to a version of the gut–brain axis that goes beyond the question of cognitive function. According to several early medical authorities, the digestive system was, at least partly, responsible for all the philosophy, music, art and literature that has ever been produced. For Galen, a pioneer in fields like anatomy, physiology, pharmacology and neurology, civilisation begins with the intestine. While the intestines of animals are smooth and straight, meaning that they must 'feed continually and as incessantly eliminate', the intricately winding coils of the human intestine permit us longer intervals between meals and bathroom breaks. And it is in those spaces between feeding and eliminating that, for Galen, culture happens. Along similar lines, the physician Avicenna, one of the great thinkers of the Islamic Golden Age and widely credited as the founder of modern medicine, argued that God had created the intestines and bladder as temporary stores for digested matter, so that we could attend to worldly matters without being continually interrupted by the need to defecate or urinate. Recast in these terms, the gut no longer appears as a distraction from intellectual business, rather it is the unlovely organ that makes it all possible. The belly, then, has not always been imagined as a hindrance to the mind, but the interactions between the two remain complex. Taking in the experiences of poets, philosophers and scholars of all stripes, this chapter explores how the life of the mind has been imagined in relation to the work of the gut.

For the psychologists at Birmingham City University, the disordered eating habits that plague the modern world are the result of mindlessly munching while scrolling through social media or staring glassily at rolling news, but long before the invention of the smartphone, physicians cautioned their patients against becoming distracted by the original hand-held technology: the book. At least as far back as the twelfth century, readers were warned against bringing books to the dinner table lest they disrupt the gut's delicate operations. By the publication of *The Anatomy of Melancholy* in 1621, the link between the act of reading and the threat of indigestion was well established, but its author Robert Burton went further to claim that the sedentary, solitary existence led by students also predisposed them to gastric disorders. A fellow of Christ Church, Oxford, who was an ordained Anglican priest and eventually a celebrated author, Burton was well acquainted with the toll that scholarly life took on body and mind. Having long suffered with dark moods, *The Anatomy of Melancholy* marked an attempt to better understand and exorcise those foul spirits; or as he put it, 'I write of melancholy, by being busy to avoid melancholy.'[121] Gathering the wisdom of Greek philosophers alongside the latest medical knowledge and his own observations, Burton concluded that melancholy was a product of both physical and mental forces. Unexplainable feelings of discontent or sadness were, for instance, said to be typically accompanied by 'rumbling in the guts, belly ake, heat in the bowel, convulsions, crudities, short winde, sowrre & sharpe belchings', which suggested some hazy connection between the intricate complexities of our emotional lives and the baser operations of the viscera.[122] Moreover, though it had roots in the ancient world, Burton feared that in seventeenth-century Britain melancholy was becoming an almost 'universall malady', an 'Epidemicall disease' from which no one was immune.[123] Those who read and wrote for a living were particularly vulnerable to low spirits. In a chapter devoted to the 'Misery of Schollers' he described how, at desks around the country, the twin evils of mental exertion and physical inactivity conspired to render their occupants depressed and gassy. Not only were scholars underpaid and typically underfed, but they were also prone to a range of debilitating 'diseases as come by oversitting' that included 'winds and indigestion'.[124] For Burton, this conspiracy of mind and gut arose from the movements of the humours, in the black bile secreted by the spleen and its negative influence on temperament, but new ways of conceptualising this vexed relationship would soon emerge.

Over the course of the seventeenth and eighteenth centuries the humoral body that we encountered in Chapter Three was slowly replaced with a body ruled over by the nervous system. This would have profound implications for how the gut was situated in culture. Beginning with the work of Thomas Willis, a London physician who credited their delicate filigree of fibres with producing all movement, sensation and thought, nerves became the object of medical and popular fascination. This was bolstered by the philosophy of prominent Enlightenment thinkers like John Locke and David Hume, who hailed the nerves as not only physiologically essential, but also necessary to our appreciation of morality, beauty, emotion and spirituality. All facets of experience, every aspect of what it means to be human, were grounded, they argued, in the subtle play of the exterior world upon the carefully calibrated strings of our nerves. The danger, of course, was that this serpentine web of sensitive fibres could be easily discomposed, either by internal tumults or external influences. How could such a finely tuned instrument endure the sensory overload of busy city streets? Could it withstand forces of commerce and competition? Would it buckle under the strain? The consensus among many physicians at the time was that the modernising world exerted too much pressure on the nerves, resulting in a proliferation of diseases like melancholy, hypochondria and hysteria in the urban populace. These nervous conditions were hard to define and their exact causes difficult to identify, but they all involved the close interplay of mind with body. Symptoms might include heart palpitations accompanied by an impending sense of doom, extreme irritability exacerbated by persistent headaches, even manic energy combined with diminished eyesight. Gut troubles were particularly prominent, with special attention paid to the possible link between disordered eating and unsettled emotions. In the new nervous body, the stomach was recast as a key site of sympathetic exchange between all the other organs, meaning that the health of the individual came to be viewed as – to a quite remarkable degree – dependent on the success of the digestive process.

Considering that medical manuals from the period listed flatulence, belching, biliousness and constipation as its most common symptoms, nervousness was a surprisingly fashionable affliction in eighteenth-century Britain and Europe. Fuelled by early self-help texts like George Cheyne's *The English Malady* (1733), which attributed them to the lure of civilisation, luxury and refinement, from the beginning nervous diseases were bound up with the aspirations of the upwardly mobile bourgeoisie (Figure 13). The lower classes were apparently unlikely to be burdened by conditions like hypochondria and melancholy because of the simplicity of their daily lives:

Figure 13: The Scottish physician and early advocate
of the meat-free diet George Cheyne. Line engraving
by James Tookey after Johan van Diest, 1787

manual labour, plain food and plenty of fresh air produced a hearty constitution, while the luxuries enjoyed
by the rich resulted in all manner of nervous complaints.
To suffer with what Robert Whytt, another prominent
doctor from the time, described as 'an uncommon delicacy
or unnatural sensibility of the nerves' was to stake out an
elevated position in the social hierarchy of the period.[125]
Physicians like Cheyne, who became a household name
after the publication of his *English Malady*, recast nervous

disease as culturally desirable and, in doing so, helped to create a thriving market of patients keen to identify with its newly flattering implications. The fashion for nervousness also elevated indigestion from a potentially embarrassing ailment to confirmation of the sufferer's superior refinement. This was achieved, in part, by the connection drawn between the excesses of a well-to-do table – loaded with '*Coffee, Tea* and *Chocolate*', with foods from around the world and dishes made irresistible by 'the ingenious mixing and compounding of *Sauces* with foreign *Spices*' – and the production of nervous symptoms.[126] But it also involved characterising the respectable stomach as uniquely sensitive, an irascible organ fine-tuned to subtle shifts in diet, environment and mood.

Beyond material wealth, nervousness could also indicate intellectual prowess; as Cheyne reassured his readers, 'It is a common Observation (and, I think, has great Probability on its Side) that *Fools, weak* or *stupid* persons, *heavy* and *dull Souls*, are seldom much troubled with Vapours or Lowness of Spirits.'[127] Rather, those engaged in the 'Arts of *Ingenuity, Invention, Study, Learning* and all the contemplative and sedentary Professions', were most likely to fall victim to the 'Diseases of Lowness and Weakness'. This was because in the nervous economy, the work of '*Study, Thinking* and *Reflecting*' wore on not only the mind, but also on the other organs and 'faculties' of the body.[128] Using similar terms, in his *Inquiry into the Nature and Origin of Mental Derangement* (1798) the Scottish physician Alexander Crichton urged readers to beware the danger of excessive study, which was likely to impair the process of digestion and bring about a 'sense of languor,

anxiety, dejection of mind, peevishness, spasmodic affections, and all the consequences of a debilitated fibre, and
[a] disordered state of nerves';[129] and earlier in the century,
Richard Blackmore's *A Treatise of the Spleen and Vapours*
(1726) claimed that those plagued by chronic gastric discomfort usually excelled 'their Neighbours in Cogitation
and all intellectual endowments'.[130] Physiological explanations for exactly why intellectual application weakened
the gut varied, with certain medical thinkers arguing that
sustained mental concentration drove blood to the brain
at the expense of the stomach, some blaming the noxious
vapours released by the half-ingested food lodged in the
gut for upsetting the mind, while others claimed that, in
fact, the trouble started in the mind, with overwrought
sentiments and unnatural excitement placing undue strain
on the delicate digestive apparatus.

Suffering scholars may have found relief in laudanum
or opium, but they would likely also have been advised
by their physicians to go on a diet. In his *Haven of Health*
(1584) Thomas Cogan, who we met in Chapter Three,
instructed students to avoid consuming 'swines flesh', as
it is a meat only suited to bodies that 'be yong, whole,
strong, occupied in labor', and instead stick to food 'which
is temperate of complexion, easie of digestion, and ingen-
dereth good bloud'. The 'weake stomack' of the scholar
was, he insisted, simply not equipped for the tough physical labour that a meat like pork demanded.[131] For Cogan,
the kind of work that the body performed in the world
defined its essential physiology and as such good health
could only be achieved by eating according to occupation.
Working in the same tradition, in eighteenth-century

Europe some physicians emphasised the value of plain foods, regular mealtimes and strict sobriety for those committed to the life of the mind. In his popular dietetic treatise *Essai sur le alimens* (1754) Anne-Charles Lorry cautioned that 'Philosophy and the lifestyle it entails are the institutions of life most contrary to nature, unless exercise, sobriety, and regularity of conduct repair their defects',[132] while the Swiss physician Samuel Auguste André David Tissot warned those engaged in literary pursuits not to try to imitate the heavy diet of the 'robust ploughman' but to stick to easily digested fare like 'light broths, niceties [and] jellies'.[133] In common with Cheyne – who put himself on a strict diet after indulging in so much fine wine and rich food that he became lethargic, melancholy, unable to walk and eventually covered in 'scorbutic ulcers' – Tissot was himself prone to debilitating bouts of indigestion that he attributed to overwork. Like self-help authors today, they sold their advice based on a dramatic narrative of personal salvation made possible by a change of diet.

Whatever the treatment or its rationale, popular physicians like Cheyne flattered their readers by drawing a link between nervous infirmity, poor digestion and genius. As the lexicographer Samuel Johnson cautioned his friend and biographer James Boswell when recommending that he read *The English Malady*: 'Do not let him teach you a foolish notion that melancholy is proof of acuteness.'[134] Despite this warning, the connection persisted and was enthusiastically endorsed by the writers, poets and philosophers of the period. In a letter to his friend Sir George Beaumont in 1806, Samuel Taylor Coleridge complained that: 'Whatever affects my Stomach, diseases

me; & my Stomach is affected [. . .] immediately – by dis-
agreeing Food, or distressing Thoughts, which make all
food disagree with me.'[135] Coleridge, who suffered from
debilitating bouts of ill health throughout his life (the
finer details of which he was notoriously keen to share
with friends, family and strangers alike), was troubled by a
litany of digestive discomforts that included constipation,
flatulence and painful wind.

Possessed by the same indigestion-inducing 'blue
devils' we met in Chapter Two, Coleridge was one of the
Romantic era's best-known sickly poets. Romanticism – a
literary, artistic and philosophical movement that reached
its zenith in the early decades of the nineteenth century
– was particularly enamoured with the figure of the sensitive genius whose intense mental activity wears upon the
health of the body. Grounded in the observations of physicians like Thomas Trotter, who claimed in his *A View of
the Nervous Temperament* (1807) that 'it is to be supposed,
that all men who possess genius, and those mental qualifications which prompt them to literary attainments and
pursuits, are endued by nature with more than usual sensibility of nervous system', illness came to be viewed as
a mark of true creativity.[136] According to historians Roy
and Dorothy Porter, the growing prominence of nervous
diseases in British and European medical discourse con-
tributed to a broader fascination with the idea of the
exceptional individual, set apart from the humdrum world
and willing to suffer for their art.[137] Romanticism, as a
movement characterised by its emphasis on authenticity,
its celebration of madness as a form of divine inspiration
and its fascination with the mysteries of the unconscious,

formed an important part of this cultural shift. Ultimately the melancholic temperament and disordered digestion that blighted Coleridge's daily life also arguably helped to secure his place among history's heroic artistic geniuses.

Excess gas and loose stools did not, however, guarantee entry into the Western literary canon, as historians James Kennaway and Jonathan Andrews have pointed out, because at the turn of the nineteenth century the 'stock of scholars in the social order was often low' and the 'relationship between bad digestion, fashion and intellectual prowess was complicated by the contrast between gentlemanly sophistication and crabbed, obsessive and unrefined academics'.[138] Distinctions between healthy and unhealthy forms of learning, worthwhile study and pointless intellectual tinkering were often made using the language of digestion. Judging whether a work of art is good or bad usually involves the question of taste, a concept that had long encompassed both aesthetic and gastronomic dimensions. In her study *Taste and Knowledge in Early Modern England* (2020), Elizabeth L. Swann points to the way that gustation, the sensation of consumption and the experience of flavour on the tongue, has helped to shape culture. Throughout the sixteenth and seventeenth centuries, she writes, the 'relationship between physical and discriminative taste' was 'intimate and overt', meaning that the processes of digestion were closely intertwined with the workings of knowledge and creativity.[139] Examples of this symbolic exchange can be found scattered through the literature of the period. In William Shakespeare's comedy *Love's Labour's Lost* (1597) Constable Dull is described as having 'never fed on the dainties that are bred in a book/

He hath not eat paper, as it were; he hath not drunk ink/ His intellect is not replenished; he is only an animal, only sensible in the duller parts', the implication being that his notoriously slow wits are the result of his reluctance to fully digest knowledge.[140] Elsewhere, a character in Edmund Spenser's epic poem *The Faerie Queene* (1590) vomits 'bookes and papers'[141] and in his *Timber; or, Discoveries Made Upon Men and Matter* (1641) Ben Jonson praises the true poet as 'Not, as a Creature, that swallowes, what it takes in, crude, raw, or indigested; but, that feedes with an Appetite, and hath Stomacke to concoct, divide and turne all into nourishment'.[142] The creation of great works of art and poetry was often likened to eating with manners and with good taste, a blending of physiological and aesthetic that positioned the work of digestion at the centre of literary life.

By the eighteenth century it was possible to not only have good taste, but also to be a 'Man of Taste' devoted to the exercise of careful discernment regarding everything from what to eat for dinner to what novels to read and what ethical ideals to cleave to. Several of these men of taste also wrote weighty treatises in which they attempted to understand the basis and meaning of aesthetic pleasure. Books like Francis Hutcheson's *An Inquiry into the Original of Our Ideas of Beauty and Virtue* (1725) and Edmund Burke's *A Philosophical Enquiry into the Origin of Our Ideas of the Sublime and the Beautiful* (1757) treated the issue of taste – what is beauty? What is the relationship between beauty and morality? How is the beauty of nature related to art? – as integral to comprehending the fundamental human condition. Taking these questions seriously meant rethinking a hierarchy of the senses that usually placed vision and hearing well above smell, taste and touch. One result of the Enlightenment obsession with taste was, according to literary historian Denise Gigante, its transformation from an 'abstract intellectual pleasure' into a 'gustatory phenomenon' that 'connoted a totality of aesthetic experience'.[143] To appreciate the subtle craft of a piece of prose or peculiar vibrance of a painting, then, one had to engage not only the eyes and the mind, but also the feelings of the gut. In an era occupied by the idea of the nervous body, a state of being defined by intense receptivity to the world and sensitivity to its effects, appetitive acts like tasting, assimilation and digestion became essential processes in the construction of the self.

The connection between reading and eating was not, then, always imagined in negative terms, rather they were often figured as metaphorically and physiologically related operations. The French philosopher Michel de Montaigne considered education and digestion to be parallel functions, while in 1597 Francis Bacon claimed that 'Some *Bookes* are to be Tasted, Others to be Swallowed, and Some Few to be Chewed and Digested'.[144] Writing in the whirl of innovation and artistic experimentation that characterised Renaissance Europe, the great scholar Erasmus described reading as a matter of ingesting and incorporating:

> The speech which moves the listener must arise from the most intimate fabric of the body [. . .] you must digest what you have consumed in varied and prolonged reading and transfer it by reflection into the veins of your mind rather than into your memory or

your notebook. Thus, your natural talent, gorged on all kinds of food will itself beget a discourse.[145]

Advising his reader not to simply commit their reading to memory, Erasmus defined true scholarship as a kind of digestive process by which knowledge is imbibed into the system. This model of active reading was essential to the humanist vision of education that had begun to emerge throughout the closing years of the fourteenth century, which sought to bring into being a new kind of citizen, eloquent, well versed in the *studia humanitatis* and ready to contribute to a wider public culture. Broad study in the humanities – in subjects like moral philosophy, poetry, history and rhetoric – was key, but most emphasis was placed on the act of reading itself and on the need to develop a discerning literary appetite. This pedagogical philosophy was grounded in classical antiquity, as Renaissance scholars looked back to the work of great thinkers like Aristotle and Plato for guidance on how to foster civic life and nurture intellectual endeavour. With these lofty goals in mind, many turned to the Roman philosopher Seneca who advised that we 'follow the example of bees'. Taking up the topic of creativity in one of his *Epistles*, Seneca observed that the lively human mind was alike to the bee moving from flower to flower gathering pollen to turn into honey. Having sifted through 'whatever we have gathered from a varied course of reading', he explained, we should then 'so blend those several flavours into one delicious compound that, even though it betrays its origin, yet it nevertheless is clearly a different thing from that whence it came'.[146] In other words, only by carefully digesting the written words of others does it become possible, somewhat counterintuitively, to create something new and truly unique. At the heart of the loftiest humanistic ideals lay the gut, grinding, dissolving and assimilating the world around it.

In his pamphlet *Areopagitica: A Speech of Mr John Milton for the Liberty of Unlicenc'd Printing, to the Parlament of England* (1644), published at the height of the English Civil War, the radical poet John Milton made an impassioned plea for the right to freedom of speech and expression. Arguing against the practice of censoring books before publication, he urged that it was only by consuming unwise or unhealthy materials that the 'discreet and judicious reader' could cultivate better judgement. 'Books are', he wrote, 'as meats and viands are; some of good, some of evil substance', but while rancid meat 'will scarce breed good nourishment in the healthiest concoction', 'bad books' encourage us to formulate counter arguments, to 'discover, to confute, to forewarn, and to illustrate'.[147] For Milton, censorship deprived the reader of the chance to employ reason and form their own view. Acknowledging such restriction as an attempt to remove the temptation of dangerous ideas and scandalous philosophies, he asked what 'wisdome can there be to choose [. . .] without the knowledge of evill?'[148] Drawing on the story of the Garden of Eden, where Adam and Eve sinned by tasting the forbidden fruit, the poet insisted that God had made such a transgression possible so that humanity might have 'freedom to choose, for reason is but choosing'.[149] Traced back to its biblical beginnings, knowledge has always been driven by appetite.

With this in mind, it is not surprising that the stomach has so often been imagined as a better route to spiritual knowledge than the eyes or the ears. As historian Helen Smith has uncovered, women's reading in the early modern period was figured as a distinctly embodied process, descriptions of which tended to draw heavily on digestive metaphors. In the margins of her copy of the *Epistles of St Paul*, for instance, Queen Elizabeth I scribbled a passage from Augustine on the alimentary pleasures of reading that reflected how: 'I walke manie times into the pleasant fields of the Holy Scripture, where I pluke up the goodlie greene herbes of sentences by pruning, eate them by reading, chawe them by musing, and laie them up at length in the hie seate of memorie by gathering them together.'[150] To imagine oneself consuming the word of God as if it were a 'goodlie' herb growing in a garden was, for the reigning monarch, to bring oneself into greater intimacy with His teachings. Having rejected the Catholic doctrine of transubstantiation, in which the bread and wine of communion are consumed as the literal body and blood of Christ, England's first Protestant queen insisted that true devotion involved a more active form of readerly gustation that required that the individual choose what or what not to eat.

While for Elizabeth I and Milton, to 'eat from the tree of knowledge' was part of what it meant to be a true and inquiring Christian, other religious thinkers were less enamoured with the idea of the virtuous stomach. In fact, many early modern writers positioned the demands of the gut in direct opposition to those of the spirit and condemned gluttonous appetites as not only intemperate, but also potentially sacrilegious. Overeaters were accused of having made a god of their belly, of having indulged in corporeal pleasures at the expense of religious devotion. Writing in 1640, the Bishop of Norwich Edward Reynolds sermonised against the 'intemperate excesse' that might lead us to make 'our belly the grave of our Soule, and the dungeon of our Reason, and let[ting] our *Intestina* as well morally as naturally farre exceed the length of the whole Man besides'.[151] Preached from the pulpit, the remarkable image of the stomach as a deathly corrupter of the soul must have severely disquieted the epicures in the congregation.

Though it is perhaps not surprising to find a minister like Reynolds denouncing bodily appetites as the enemies of reason – his Presbyterian theology would have left him especially attuned to the danger of earthly pleasures – a similar antagonism also emerged in less expected places. According to historian Anne C. Vila, even in Enlightenment France, where 'social pleasures like fine dining were central to the effort to redefine the modern intellectual as a public-spirited, convivial fellow eager to partake of worldly life', anxieties over the consequences of such 'belly-centred' excesses remained.[152] Alongside the image of the scholar as a gourmand, whose refined culinary tastes were rightly bound up with their intellectual endeavours, sat the less attractive figure of the sickly writer we encountered in Chapter Two, who was beset by digestive troubles and creatively constipated. Success in the literary world of cities like Paris meant treading a fine line between gustatory enjoyment – attending the right parties, securing a spot next to a wealthy patron at the dinner table and so on – and the kind of gluttonous overconsumption likely to impede the higher functioning of the mind. Danger also lay in too readily conflating the pleasures of the gut with the work of the mind. As Coleridge protested in his *Philosophical Lectures* (1819), the analogy between literary and gustatory taste had been taken so far that now 'one man may say I delight in Milton and Shakespeare more than Turtle or Venison' and another that a dish of 'turtle and a good bottle of port' is preferable to any poetry, but 'you must not dispute about tastes'.[153] Bind creativity too closely to the work of the body and risk turning great literature into little more than another dish on the table to be sampled or passed on in favour of another.

Coleridge's fretting at the denigration of poetry as consumable like 'Venison' reveals the underlying tension between the work of the gut and the work of the mind to which this chapter has addressed itself. Digestion and thinking have often been set in opposition to one another – the suffering of the scholar whose excessive study brings on bouts of indigestion or the bibliophile warned against reading at the dinner table – but the perception of some connection between the two has allowed the gut to occupy a surprisingly prominent place in the history of Western culture. Following the movement of meaning from victuals to vitals, writers, philosophers and poets explored the physiological and aesthetic dimensions of taste in a way that transformed the alimentary canal into an arbiter of beauty, morality and truth. Taken up as popular metaphors for learning, from the Renaissance onwards acts of consumption and digestion became key to the imagination of an educated and well-rounded, modern citizen. Bumping up against this ideal, the discourse of nerves that emerged over the eighteenth century pitted the sensitive artist or scholar against a booming urban world that threatened to overwhelm their delicate, finely attuned facilities. Moving into the nineteenth century, the next chapter picks up on the idea of a nervous economy in which the seemingly unceasing demands of a burgeoning modernity threatened to deplete the body's finite energy reserves, stymieing thought and hindering digestion in the whirl of industrialisation, urbanisation and commercialisation. Here attention lies, however, not with the romantic distresses of brilliant poets and the musings of learned philosophers, but with the rather more mundane realm of the office and the question of what to eat for lunch.

6

Eating at the Desk

Sandwiches have ruined lunch. According to the *Daily Mail*, the invention of the pre-packed sandwich, sometime in the early 1980s, marked the end of convivial midday dining and ushered in an era of eating on the move.[154] For its detractors, the ubiquity and tempting convenience of the sandwich has succeeded in rendering the meal a rather joyless affair. Given that most of us eat lunch as part of a working day, such criticisms are often bound up with broader concerns over the demands and stresses of modern office culture. Purchased between meetings and eaten in front of a computer, these triangles of doom are held up as an everyday example of our dangerously skewed work/life balance. In 2016 the *New York Times Magazine* ran a series of portraits by the photographer Brian Finke that captured the horrors of desktop dining: a glassy-eyed community manager slurps soup in a staff meeting, a television producer types with a slice of pepperoni pizza hanging from her mouth, a lawyer munches crisps over a pile of legal documents and the desk of a city trader is strewn with food wrappers, napkins and not one, not two, but six bottles of hot sauce.[155] In the article that accompanied these images, eating at the desk was presented as a uniquely North American phenomenon, one driven by a work culture in which staying late at the office is expected and stopping for lunch is viewed as a sign of idleness. Desktop dining is, however, on the rise around the world: *The Times of India* recently advised its readers on olfactory etiquette (vegetable roti and yogurt are fine, hard-boiled eggs and goat trotter soup should be avoided), Norwegians arrive to work armed with a *matpakke* – a rather uninspiring open sandwich of cheese, meat or jam – and when asked, almost 70 per cent of South Africans admitted to working through lunch most days.[156] Even in southern European countries like Spain and Italy where the working day has traditionally been structured around breaking for a long lunch – with multiple courses and often a nap to follow – things are changing. During the Covid-19 pandemic the French government was forced to repeal a long-standing ban on desktop dining. Previously, the *Code du travail* had explicitly prohibited 'allowing workers to have their meals in places dedicated to work', but the closure of restaurants and social distancing measures meant that the regulation was no longer enforceable.[157] For many, this reversal marked not only the desecration of French food culture, but also the further encroachment of work into life. What better measure of the increasing pressure to perform better and produce more than our inability to stop – even briefly – to eat lunch? Such grumblings are usually accompanied by nostalgia for a world before the tyranny of the sandwich, when food was prepared with care and time taken to enjoy it. Which is all very nice, but such hazy imaginings are hard to locate in the past.

After all, lunch has long been inseparable from labour, a product of the working day rather than an escape from it.

Lunch was a Victorian innovation. A response to lengthening commutes and longer hours, in Britain it was a product of the rhythms and peculiarities of the working day. In medieval times most people ate only two meals a day, a large dinner served in the late morning or early afternoon and a small supper eaten in the evening, and it took until the seventeenth century for breakfast to become part of the daily routine.[158] Something called 'nuncheon' appeared in the middle of the fourteenth century and the term 'luncheon' began circulating in the seventeenth. But neither denoted a mealtime as such, only a light repast: in his 1755 *Dictionary of the English Language* Samuel Johnson suggested that it meant simply 'as much food as one's hand can hold' and when 'lunch' first appeared in *Webster's Dictionary* in 1817 it was used to describe simply a 'large piece of food'.[159] Only with the cataclysm of industrialisation and the transformations enacted by urbanisation did lunch emerge as an essential break in the working day. It is telling that familiar conjunctions like 'lunchtime' and 'lunch hour' entered common usage around the middle of the nineteenth century, just as new patterns of work and travel were recalibrating the rhythms of everyday life. Prior to the institution of Greenwich Mean Time (GMT), which was first adopted by the railways in 1847 to regulate train timetables before being publicly inaugurated with the construction of Big Ben in 1859, most people looked to local markers of time to structure their day: the sundial in the middle of the village square, the chiming of church bells, labourers returning from the fields, a candle in the parlour burned down to the wick. Industry, like the railways, required standardised time to function. Reflecting on this shift, the historian E. P. Thompson has described how the 'first generation of factory workers were taught by their masters the importance of time'; in other words, they had to develop a new sense of time, to reset their internal clock to the demands of a new economic order.[160] The precise division of time on the factory floor extended beyond the production line to the body of the worker, not only orchestrating the movement of their limbs but also – by instituting a set hour for the consumption of lunch – establishing the temporal cadence of the digestive system.

The dazzling ascent of lunch through the middle of the nineteenth century created a booming market for daytime eating. Walk around London, Manchester or Glasgow in the early afternoon and you would encounter a thriving dining culture: from street vendors and chop houses to vegetarian restaurants, cheap cafes to exclusive members' clubs, lunch was woven deep into the fabric of the Victorian city. But outside of the restaurant industry, the advent of this new meal was not met with universal enthusiasm. Instead, in common with the rise of the pre-packed sandwich today, the advance of lunch provoked anxieties over the health of the nation and the quickening pace of urban life. Described by a witty commentator as 'an insult to one's breakfast and an outrage to one's dinner', lunch was seen as a disruption of man's natural eating schedule: breakfast was taken too early, the evening meal was indefinitely postponed, and workers were forced to fill the gap with ill-considered and hastily consumed foods.[161] Where once it had been possible to make it back home to eat a proper meal, at a solid table surrounded by a loving family, the speed of modern life had rendered that simple pleasure impossible. Industrial capitalism was, according to a host of worried commentators, a disaster for the stomach, overtaxing and unbalancing its delicate operations, so that the nation's workforce became exposed to a host of debilitating gastric complaints.

Nowhere were the dangers of eating on the go more obvious than on the railways. With the vast expansion of the rail network and advances in locomotive technology through the century, train travel became a feature of working life for an increasing number of daily commuters and those travelling for business. The passenger forced to wolf down breakfast to make the morning train, bolt a week-old meat pie in the grubby station refreshment room or, worst of all, attempt to eat a sandwich on board as the train raced recklessly along the track, quickly became the focus of medical concerns over the fate of the stomach in a world that seemed to be speeding up. The problem with eating on the move was, according to the dramatically titled *Hurried to Death; or, a Few Words of Advice on the Danger of Hurry and Excitement, Especially Addressed to Railway Travellers* (1868), that it placed undue stress on the digestive system. As the pamphlet's author Alfred Haviland warned, 'thoughtless hurry and exertion' on a full stomach overtaxed other organs of the body like the heart and could lead to 'premature disease and untimely death'.[162] A rushed lunch could, it appears, have potentially fatal consequences.

By the 1880s, the situation had apparently deteriorated

to such a degree that the magazine *All the Year Round* reported that dining rooms catering exclusively to dyspeptics were 'soon to be opened in London' where 'a doctor would examine each person' as he or she entered and then 'prescribe' a lunch menu best suited to their unique dietary needs.[163] Though the craze for physician-led dining does not appear to have taken off, the notion that there might be money to be made from the upset stomachs of the capital's workforce speaks to the prominence of gastric trouble in the national imaginary. The anxiety that seems to have been provoked by lunch can be, at least partly, attributed to the Victorians' obsession with the health and efficiency of the working body. For medical professionals, policymakers and industry leaders, urban lifestyles and unhealthy diets were putting the nation at risk of physical degeneration. In this climate, a well-managed body came to constitute an essential aspect of civic virtue, one that signalled the individual's commitment to the economic success of the country and the strength of the empire. The question of how to cultivate efficient labouring bodies was most notably taken up by dietary reformers, who framed their analysis of the digestive system in the utilitarian language of supply and demand, investment and return, debt and repayment, excess and waste. Popular dietetics, published in dedicated tracts and disseminated in middle-brow periodicals, offered one way of negotiating the demands of the irascible gut. Presented as hybrid texts, composed of treatises outlining the functioning of different parts of the digestive system and lifestyle guides advising on diet and physical exercise and warning against the cultivation of bad habits, literature on nutrition occupied a space between medical science and 'self-help'. This emerging scientific discipline formed itself in relation to the world of work, by emphasising the need to balance resources consumed with energy expended, and by structuring its analysis of fats, proteins and carbohydrates around the needs of different professions.

As well as featuring tables depicting the daily calorific requirements of workers in a broad range of sectors, nineteenth-century literature on nutrition also warned against the dangers posed by certain kinds of employment. Those engaged in white-collar occupations, clerks, writers and office workers of all kinds, were seen to be especially vulnerable to gastric disorder. Beginning with the publication of W. M. Wallace's *Treatise on Desk Diseases* (1826), which claimed that adopting a 'stooping position at the desk' placed 'injurious pressure' on the stomach, the office was charged with inflicting all manner of digestive complaints.[164] While manual labour was widely praised as stimulating appetite and aiding healthful digestion, life in the office threatened not only dyspepsia but also nervousness. Poor dietary habits – long days fuelled by dry biscuits secreted in desk drawers, cheap pies purchased in haste from disreputable street vendors, overdone chops and ruinous liquor – were compounded by the sedentary nature of the work, the sustained mental concentration demanded by ledgers and account books, exacerbated by the strain of lengthening commutes and the precarious financial situation endured by those towards the bottom of the ladder. As the food historian Andrea Broomfield has noted, many lower-middle-class workers were badly paid, and this was evident in the rather paltry 'slice of cheese and bread' – perhaps a precursor to today's dreaded pre-packed sandwich – that was common to most 'desks around noon-time'.[165] Of course, manual labourers were even worse off as most relied on street traders selling cheap portable fare like sausages that were sometimes made with rotting meat and eel pies that were prepared with flour bulked out with plaster dust. Yet white-collar workers remained the primary focus of cultural concern. Imagined as a victim of the growing city, where there was, as one dietary reformer had it, 'more study anxiety, exhausting pleasure, haste to become rich, exciting and often distressing news, than in any previous age', the dyspeptic clerk embodied an emerging sense of disquiet over the way the world was going.[166] In the whirlwind of working life and the fevered excitement of the urban world, indigestion seemed almost unavoidable.

The triangulation of work, mental strain and indigestion fuelled both factual and fictional depictions of life in the Victorian city. In George Gissing's *New Grub Street* (1891), a novel obsessed with the perils of working life, the thwarted professional ambitions of its characters are expressed most clearly in their various gastric complaints. Described as 'martyrs to dyspepsia', Gissing's hack journalists and failed writers find themselves frequently without appetite and plagued by debilitating bouts of nausea; the results, we are told, of 'toiling' unsuccessfully in an increasingly competitive marketplace.[167] Joining the alimentary and the environmental, indigestion in *New Grub Street* is not only a physical response to poor diet, but also a mental state shaped by the increasingly precarious nature of modern employment. A committed chronicler of lower-middle-class life, in novels like *Eve's Ransom* (1895), *The Paying Guest* (1895) and *The Town Traveller* (1898) Gissing dramatised the experiences, interests and concerns of a growing demographic of white-collar workers. Taken up elsewhere by writers like Arnold Bennett, Walter Besant and E. M. Forster, by the end of the nineteenth century the life of the office-dweller had been established as a worthy subject of fictional enquiry. The ubiquity of the clerk in turn-of-the-century culture was due, at least in part, to the extraordinary rise in their numbers over this period, from 80,109 men and 3,101 women in 1881 to 140,847 men and 39,847 women in 1911 in London alone.[168] The capital transformed itself to meet the needs of this growing army of desk workers. Suburbs cropped up around the edges of the city spawning networks of railways, tram lines and underground trains. While in town, new entertainments – department stores, skating rinks, tea houses, sports clubs and parks – flourished on the small disposable income now afforded to the average middle-class family.

The intimacy of clerk and urban environment transformed not only the city, but also the working body. *New Grub Street*'s cast of sickly writers were drawn from cultural representations of the unhealthy office-dweller, made ill by the pressures of the modern workplace, where wages were stagnating, job stability was under threat, advancing technologies threatened extinction and an influx of female workers promised to lower the status of the profession. Though there is little consensus among historians regarding the severity of this crisis, popular depictions from the period tended to present the life of the office worker as chiefly characterised by insecurity, financial strain and monotony. Fears that urban life might prove antithetical to physical and mental health coalesced around the figure of the clerk, whose sedentary and disordered habits rendered them especially vulnerable to degeneration. As a witty commentator asserted in an 1890 newspaper article, 'the dyspeptic' should be hailed 'a victim of civilisation, a martyr to the times, a sacrifice to the *fin de siècle*, a heart-burned offering to the Modern Spirit'. He is in fact, the article continued, 'the new Prometheus [but] his offence was not to steal fire from the gods. It is not deemed sacrilegious – on weekdays – to bolt one's breakfast, to race for trains, to omit one's lunch, to walk only when there is a bus strike on.'[169] Such caricatures did not necessarily reflect the reality of office life, but they did speak to broader medical and socio-economic concerns over the health of the working body.

Ironically, the clerk, overworked, riven by chronic digestive problems, plagued by debilitating anxiety and victimised by the relentless pace of city life, was also a highly valued customer. Capitalism was the cause of gastric distress, but it also offered solutions in the form of patent medicines like Fennings' Indigestion Tablets, Huxley's Ner-Vigor and Chologestion, which promised to free the consumer from the burden of ill health and ready them for the challenges of the modern world. In an advertisement for Burdock Blood Bitters, an American product that claimed to address everything from constipation to kidney complaints and nervousness to so-called 'female weakness', we see a well-dressed businessman order – much to the astonishment of the hotel concierge – pie for 'breakfast, dinner and supper'. He can do so with confidence and also roll the dice on the odd 'rail road sandwich', because Burdock Blood Bitters keeps his 'stomach and digestive apparatus in perfect order' (Figure 14). Tonics, tablets and all manner of cures were sold to the urban employees as simple solutions to the problem of overwork, as miracle products that would help the canny consumer to withstand the pressures of the office.

It was not only purveyors of quack medicine who recognised the stressed-out desk worker as potential customer. Beginning in the 1880s, advocates of the vegetarian diet began to foreground the specific health benefits that meat-free dining held for those employed in sedentary occupations. Publications like *The Dietetic Reformer* and *The Herald of Health* featured editorials warning of the deleterious effects of flesh eating on the health of the brain worker and published articles celebrating the transformative power of legumes on productivity. They also ran letters that boasted of conversion and redemption. Writing to *The Vegetarian*, a 'City Clerk' reported that as 'a flesh and white bread eater' he had been 'afflicted with that great evil, constipation, accompanied by depression', but that these ailments had disappeared with a change of diet.[170] When, towards the close of the century, vegetarian restaurants began springing up in Glasgow, Manchester and London, they marketed themselves directly to the city's workforce by offering lunchtime deals and emphasising the digestibility of their light, fleshless fare. Walk into a vegetarian restaurant in Farringdon or Soho in 1890 and you may have encountered a meeting of the Vegetarian Society or heard a barnstorming speech from an anti-vivisectionist, but you would be far more likely to be met with

Figure 14: Advertisement by Foster, Milburn
& Co. for Burdock Blood Bitters, 1880

tables of hurried workers wolfing their midday meal. As a
contributor to the *Vegetarian Messenger* noted in 1887, the
restaurants around Cheapside had become popular with

low-paid clerical workers: the Alpha on Oxford Street was used by staff from the nearby Crosse and Blackwell building, the Ceres near St Paul's was popular with workers from the surrounding warehouses and elsewhere a newspaper reporter remarked with surprise that one meat-free establishment was 'crowded with customers, many of whom were evidently substantial City men'.[171] With names like The Fig Tree and The Garden, they held out the promise of wholesome, nourishing fare in the heart of the city, a taste of nature in a bustling urban environment and a chance to escape briefly from the unceasing demands of the desk.

The human digestive system was, it appeared, simply not equipped to deal with the demands of omnibuses, electric light, telegrams and mass-produced meals. And then, as now, the rushed and uninspiring lunch served as a means of articulating dissatisfaction with the pressured and precarious conditions of working life. Importantly, though the discourse of nutritional science played a disciplinary role by attempting to create cost-effective productive bodies, throughout the nineteenth and into the twentieth century it also provided workers with a new language with which to protest economic uncertainty and the unfair distribution of resources. Writing on the Coronation Bus Strike of 1937, in which 27,000 London bus workers walked out for better working conditions and a seven-and-a-half-hour day, historian Rhodri Hayward has uncovered the gastroenterological factors at play in this industrial dispute.[172] Throughout the early 1930s more and more drivers fell ill with what came to be known as 'busman's stomach', a form of painful gastritis said to result from peculiar physical and psychological pressures of the job: following a tight timetable, drivers ate hurried packed lunches on the move, often while breathing in harmful carbon monoxide fumes leaking from the engine into the cabin. The growing prevalence of this digestive disturbance in the workforce became a key bargaining chip in the industrial dispute of 1937; embodied evidence, identified by doctors and cited by union organisers, of the unmanageable pressure that drivers were being placed under.

This was not, moreover, the first time that squiffy stomachs had played a mediating role in negotiations between workers and management. In the early years of the twentieth century post office clerks around Britain threatened to strike over the epidemic of gastric disorders plaguing their profession. These were the result, according to an article in *The Postal Clerks' Herald*, of both the hunched posture adopted by workers at their ledgers which caused 'intestinal stasis and the misery of constipation' and the practice of snacking on dry crackers throughout the day instead of stopping for lunch, a foolish habit that posed a threat to health, as well as affording clear evidence of overwork. Lunch, then, was understood not only as a problem, but also sometimes as a potential solution to the outbreak of gastroenterological distress threatening the nation's productivity. The union demanded the institution of dedicated lunch hours and the introduction of canteens, as well as reduced working hours, and used the upset stomach as a key tool in their negotiations. In response, in 1912 the government instituted a formal inquiry into the physical and mental health of postal employees, which heard from dozens of witnesses and included testimony from several medical professionals. Asked to explain the notable prevalence of nervous complaints among mail sorters, one expert suggested that the nature of the work itself – which was repetitive yet demanded sustained concentration – was to blame. Sorters, he claimed, were in a state of 'continual worry' and this mental strain resulted in debilitating digestive ailments.[173] The gassy burbles and deep grumbles of the stomach were, it seemed, a kind of unconscious protest at the stresses of the modern workplace.

One solution to the psychological and physical problems raised by these industrial disputes lay with the workplace canteen. Throughout the second half of the nineteenth century, the Factory Inspectorate – a regulatory body established to monitor safety in the workplace – urged employers around the country to introduce cafeterias. It was essential, the inspectorate urged, that factory workers be provided a dedicated time and space to eat away from dangerous industrial pollutants. Despite the regulatory body's repeated calls to action, most factory owners were reluctant to take on the expense of onsite catering facilities and remained unconvinced of the apparent benefits. Part of the problem, as historian Sue Zemka points out in her study *Time and the Moment in Victorian Literature and Society* (2011), was that factory owners did not view lunch in terms of its benefits to health or the opportunities for sociality it might provide, but essentially as a waste of time. Throughout most of the nineteenth century, industrial productivity was, Zemka writes, based on the 'cunning robbery of small increments of time. Ten minutes more to begin the working day; a quarter hour off of lunch; another half or quarter hour added at the end of the day', an unforgiving pace that certainly did not allow space to pause, sit and eat.[174] Only with the outbreak of the First World War did the factory canteen finally take off in Britain, with the numbers of dedicated facilities rising from around 100 before 1914 to over 1,000 by 1918. As historian Vicky Long has uncovered, this ten-fold increase was due to the demands of munitions factories, where a new labour force, composed largely of teenage boys and women, was being called upon to maintain the supply of weapons to the front line by working long hours in dangerous conditions. During this period, overwork and fatigue among factory workers came to be viewed as matters of national importance; issues that might, if left unaddressed, imperil the war effort. This resulted in what Long has described as 'an unprecedented interest in improving workers' health' that saw far greater attention paid to 'physiological needs' and 'in particular their dietary requirements'.[175] According to a government committee report from 1917, factory canteens were essential to maintaining productivity because 'the efficiency of the manual worker depends as directly upon his food supply as does the mileage of a motor car upon its petrol'.[176] Likening the labouring body to a machine, one that required rest and regular refuelling, the canteen came to be seen as intrinsic to the operations of the factory. Concern for the dietary regimen of employees remained, in this case, tightly bound to issues of capitalist production and national efficiency, rather than to any broader sense of collective wellness or individual fulfilment.

Alternative visions of the factory canteen could be found, however, in planned industrial communities like

New Lanark, Port Sunlight and Bournville. Social reformers like the Welsh philanthropist Robert Owen, who in the early nineteenth century established a cotton mill to the south-east of Glasgow and a fully functioning town to service it, were invested in the welfare of workers beyond the walls of the factory. Often cited as a progenitor of the cooperative movement, New Lanark was an experiment in utopian socialism that provided textile workers with well-designed housing, a school for their children and an educational institute where they could continue their own studies. Owen also made provisions for a large kitchen and canteen, where the community could come together every evening to eat. As well as ensuring that everyone was well fed, the experience of collective dining provided opportunities to socialise and relax after the rigors of the day. Later in the century, when Quaker brothers George and Richard Cadbury decided to expand their chocolate-making operation, they looked to the paternalistic example set by Owen and established a model village for its workers. Located in the countryside outside of Birmingham, Bournville was built with the health of its residents in mind. Along with several large parks, the Cadburys also laid football pitches, running tracks and bowling greens, established a fishing lake and even installed an outdoor lido. There were no public houses, but once again emphasis was placed on the need to provide workers with places to eat and plans for the town included several staff canteens (Figure 15). Both Owen and the Cadburys remained invested in the profit-making aspects of their businesses, but they also recognised the role that lunch played in the creation of a healthy, happy, more effective labour force.

Figure 15: The staff canteen and kitchen at
Bournville, early twentieth century

Today, some employers – notably tech giants like Meta
and Google – are following the example set by these nineteenth-century reformers and providing their staff with
impressive onsite dining options. Lunch at Apple's Cali-
fornian headquarters might be a freshly baked sourdough
pizza or a hearty bowl of ramen, healthy sushi or an indul-
gent burger, fish tacos or $1 oysters, all heavily subsidised,
cooked by some of the country's best chefs and consumed in
a light-filled room surrounded by nature. Not all businesses
are able to offer their employees such a pleasurable lunchtime experience, but more are now looking to the canteen
to improve morale, foster stronger 'company culture' and
boost productivity. The degree to which access to a salad
bar impacts how you feel about your job – more than, say,

higher wages or reduced working hours – remains to be seen, but the recent resurrection of the cafeteria suggests the question of what to eat for lunch and where remains key to understanding our complicated relationship with work.

The history of lunch, a meal born of Victorian industrialisation and urbanisation, still speaks to many of our own anxieties over how to eat, work and live well in the modern world. From the beginning, the consumption of lunch was bound up with the demands of the workplace and the hurried midday meal came to exemplify the pressures it exacted on the individual. Pressures that made themselves known in overstrung nerves and gassy stomachs. The construction of lunch as in some way inherently problematic, as inevitably a site of anxiety and ill health, might also – as we saw regarding the threatened postal strike – be read as a discursive strategy, as a way of voicing discontent at the precarious and pressured conditions of the working day. All of which suggests that there is rather more to the humble prepacked sandwich than meets the eye. Work is, as the last three chapters have argued, essential to understanding the history of the gut. Connecting the individual body to the working world, the digestive system labours internally and is entangled with commutes, offices and factories. Lunch is, as this chapter has explored, one of the most obvious sites in which stomach and work become entangled with one another, where the rhythms of the gut meet the structures imposed by capitalism. This is, at least in part, a matter of temporality, where the peculiar cadences of the belly meet the carefully scheduled working day. Taking up the question of time in more depth, the following chapters consider the past, present and future of the gut.

TIME

In 1928 a distinguished professor at Johns Hopkins University published a bestselling guide to child rearing, *Psychological Care of Infant and Child*, that encouraged parents to treat their young offspring as 'if they were adults', to avoid excessive physical affection and to operate a strict unchanging daily timetable.[177] Outlined in detail for the reader, this schedule dictated exactly when children should be woken, what they should eat for breakfast, how much time should be allocated to playful romping (7.30 a.m. to 8.00 a.m.), when bath time should occur and when it was time for bed. Its author, a behavioural psychologist by the name of John B. Watson, held that most responses to the world are learned rather than innate or inherited and that, as such, our early experiences play a definitive role in shaping what kind of person we will grow into. Parents should, he urged, devote themselves to instilling good habits in their children from an early age and nowhere was this more important than in the bathroom. As part of his prescribed daily routine, from the age of eight months the child should be put on a 'special toilet seat into which he can be safely strapped' and then left alone until they defecate. Isolation was key, the child should be 'left in the bathroom without toys and with the door closed', and it was advised that they remain there for up to twenty minutes or until the bowel movement was completed.[178] The well-trained child was produced through regularity, an imperative that Watson extended to the functioning of the gut.

Today we might doubt the wisdom of dictatorial methods that discouraged parental tenderness in favour of lashing infants to toilets, but potty training still necessitates a kind of time management. Though the age at which children typically transition out of nappies has changed, it remains pivotal to infant development as a moment of entry into the rules, conventions and expectations of the wider world. It involves learning to exercise control over previously unconscious processes and, importantly, to measure and portion time appropriately. As Jean Walton puts it in *Dissident Gut: Technologies of Regularity, Politics of Revolt* (2024), 'toilet training involves nothing less than the mastery of the bodily regulation of time: the division of duration into evenly spaced evacuations of the large intestine, the measuring of progress by rhythmic contractions of smooth muscle'. The temporal disciplining of the bowel integrates the child into society by distinguishing them from the 'indiscriminately excreting animal' and affiliating them with the 'systematically excreting human being'.[179] Recalling Galen's attribution of all human culture to intricate coiling intestines that allow us, unlike animals with their short innards, time between eating and eliminating to think, make and create, toilet training has been similarly imagined as a way of civilising the body.

That there is a temporal dimension to this process

– the unruly cadences of the child's large intestine forced to keep pace with the exacting schedules of nursery, school and eventually work – speaks to the persistence of time as a key concern in the cultural history of the gut. Think of William Beaumont carefully recording how long a piece of chicken took to dissolve in the exposed stomach of his unwilling patient or the Victorian workers forced to align their digestive rhythms with the standardised tempo of the factory floor, the operations of the gut are entangled with time in many different ways. The following chapters weave together sewers, entrails and alimentary tracts to explore the belly's temporal dimensions. Paying particular attention to bodily waste, how it has been utilised, managed, avoided, celebrated and theorised, they argue that the binding of digestion to time has had profound implications for how we understand our bodies and their place in the world.

7

Present

In 1851 London's Hyde Park played host to the Great Exhibition, a celebration of culture and industry showcasing technological innovations from around the globe. Alongside printing machines, steam engines and the latest communication devices, visitors to the exhibition could 'spend a penny' to use the world's first public flushing lavatories. Installed in the Retiring Rooms of the Crystal Palace, they were the work of George Jennings, a plumber based out of a small shop off Blackfriars Road in Southwark (Figure 16). More than a convenience, the toilets were a popular attraction, hailed in the press as an engineering triumph and cited as proof that Britain was leading the way when it came to the disposal of human excrement. In the World Fairs that followed, hosted by cities including New York, Chicago and Brussels, sanitation technologies were also prominently featured: the International Health Exhibition of 1884 held numerous displays of drainpipes, visitors to the 1867 Exposition Universelle in Paris were invited on guided tours of the city's impressive sewer system and Dresden's 1911 International Hygiene Exhibition showcased new architectural models that were set to revolutionise Germany's sewage disposal plants. It is significant

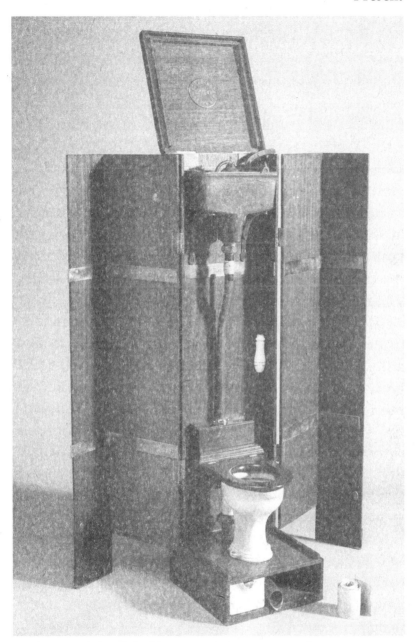

Figure 16: Model of Jennings' patent water closet, c.1900

that in these grand exhibitions, where the spoils of empire were ostentatiously arrayed and the myriad achievements of the host country were broadcast to the rest of the world, the question of how best to dispose of human excrement should have demanded quite so much attention. Part of what it reflects is the belief, widely held at the time, that social and cultural progress could be best measured in sanitary terms. The nineteenth century was, among other things, the age of sanitary science, a discipline described in 1857 as that 'which deals with the preservation of health and prevention of disease in reference to the entire community, as contradistinguished from medical science [...] which has for its aim the restoration of health when lost'.[180] Encompassing a wide range of individual and collective actions from improved personal hygiene to mass vaccination programmes, better-ventilated homes to safer food practices, at the heart of this scientific revolution was the question of how to collect and safely dispose of the waste produced by bodies thronging the streets of booming cities and industrial towns. Toilets were proudly showcased at the Great Exhibition because the onward march of civilisation appeared to rest on the successful management of waste.

To Victorians perched on its generous wooden seat, the flushing toilet was an astounding feat of modern engineering and ingenious design that encapsulated the unique drive and ambition of the age. It was an invention that seemed to speak directly to the present and in their enthusiastic advocacy of the toilet's transformational potential, sanitary scientists elevated the gut as an arbiter of technological and medical progress. As we have already seen, throughout history the stomach – with its endless demands, pleasures and mysterious disorders – has served as a rich source of metaphor and a material site where social, economic and political conditions could be made manifest. This commitment to the organ's storytelling power has, however, been typically accompanied by a countervailing impulse to distance the present moment from the digestive process and its more unappealing products. Another way to think about the temporality of the gut is to consider how the present has been defined against the excretory practices of the past. Conceptions of what it is to be 'civilised' or 'modern' have, across time and in different contexts, been forged in direct opposition to the perceived filth of the past. Focusing its attention on nineteenth-century London and Paris, this chapter will argue that how we think about the products of digestion reveals a great deal about how we have imagined ourselves and the kind of society that we want to live in.

Visitors to the Great Exhibition in 1851 might have been forgiven for doubting whether London was quite the magnificent centre of sanitary reform that it claimed to be. Throughout most of its history the city had used the River Thames as a watery dumping ground for human, animal and industrial waste. As its population grew, bursting with newcomers from the countryside, workers from all over Britain and immigrants from around the world, the river became increasingly polluted. Writing in 1855, the scientist Michael Faraday described peering down into the water and observing 'feculence rolled up in clouds so dense that they were visible at the surface'[181] and the Tory politician Benjamin Disraeli characterised it in Parliament as a 'Stygian pool reeking with ineffable and unbearable horror'.[182] Not only was the Thames a national embarrassment, a stinking cesspit at the heart of the capital, it was also feared as a source of illness. Between 1832 and 1866 the city was devasted by four major outbreaks of cholera, a terrifying disease that induces stomach cramps, diarrhoea, vomiting and severe dehydration in those it infects. Writing as the first epidemic was gathering pace, a reporter for the London Gazette described the distinct physical appearance of sufferers: 'the eye sinks, the look is expressive of terror and wildness [. . .] The skin is deadly cold and often damp, the tongue always moist, often white and loaded, but flabby and chilled, like a piece of dead flesh.'[183] Today we know that these potentially fatal symptoms are caused by the presence of a particularly nasty bacterium, Vibrio cholerae, which is usually introduced into the small intestine through contaminated drinking water. But in the nineteenth century its origins were still a mystery. One idea posited that foul air, generated by rotting vegetables, decaying animal matter and putrefying human waste, was the cause and many looked to the river as the likely source of the noxious vapours. This explanation was grounded in the theory of miasma, the origins of which can be traced back to the Roman Empire, which held that bad smells, lingering in the atmosphere and poisoning the air, were responsible for causing illness. Though the miasmic hypothesis was eventually debunked, it is understandable that the horrible stink given off by the clouds of 'feculence' that Faraday observed roiling in the Thames could provoke disgust so visceral that it might bring with it the fear of death.

The malodorous issue was further exacerbated by the waste produced by new industries springing up along its banks and by the installing of household toilets that flushed directly into the river. Rather than precepting a sanitary revolution, these early domestic toilets upset the digestive system of the city. Until well into the nineteenth century, the disposal of human waste was primarily taken care of by a group of workers known as 'night soil men'. Journeying through the darkened streets, they would collect the city's ordure from communal cesspools and transport it to the surrounding countryside, where it would be sold on to farmers who would use it to fertilise their crops, which would then be shipped back to the city to feed its inhabitants. The increasing use of the river as a dumping ground for the excreta of the growing metropolis presented, according to the social reformer Henry Mayhew, two major problems. For one, it represented a grave mismanagement of the faeces that could be being used as fertiliser, as he estimated that with a return of £10 for each 100 tons of sewage Londoners were 'positively wasting £4,000,000 of money every year' or, put another way, flinging an annual '246,000,000 pounds of bread' into the Thames.[184] Profligacy that, in turn, had polluted the drinking water, disrupting the proper circulation and transformation of waste material. So that, as Mayhew complained, the 'water in which we boil our vegetables and our meat' is 'impregnated over and over again with our own animal offal [. . .] We drink a solution of our own faeces.'[185] Without the intervention of the night soil man and the farmer, who worked together to transform waste into food that could be reincorporated back into the ecosystem of the city, sewage became a source of toxicity, contamination and disgust.

The situation in London came to a head in the hot summer of 1858, when exceptionally high temperatures and an extended period of dry weather resulted in a stench so rotten, so foul, that it threatened the very business of government. During what came to be known as the 'Great Stink' the Houses of Parliament were engulfed by a cloud of noxious gases and its members – many of whom had been previously reluctant to undertake the expensive and disruptive work needed to solve the malodorous problem – were forced to act. In a record eighteen days, a bill was written and passed that gave the go-ahead to an ambitious scheme to install a vast network of sewers under the streets of the capital. The architect of the project was Joseph Bazalgette, a civil engineer who planned and oversaw the construction of 1,300 miles of brick-built street sewers, as well as five pumping stations and several embankments along the river. Finally completed in 1875, Bazalgette's system cleaned up the Thames and radically transformed the olfactory experience of the city.

The 'Great Stink' was widely condemned as a national disgrace, but London was by no means alone in its problem with waste. During what historian David S. Barnes has described as the 'Great Stink of 1880', Paris was gripped by a putrid honk emanating from overburdened sewage treatment plants.[186] Residents were outraged, newspapers carried headlines condemning the government's response to the crisis and scientists warned that the noxious atmosphere posed a grave threat to public health. This smelly episode was not the first or even the worst in the city's

history, yet the anger, fear and disgust it prompted out-
stripped anything that had come before. What was it that
made this particular stench so intolerable? One answer
might lie in the complaint voiced by the popular French
publication *Le Siècle* that 'our great and beautiful city'
was being turned into 'an immense cesspool'.[187] In other
words, the fug of excrement hanging over the streets of
Paris imperilled not only the wellbeing of its individual
residents, but also the reputation of the city as the apex of
European culture. Civilisation and shit do not, apparently,
mix. In his famous history of smell, *The Foul and the Fragrant: Odor and the French Social Imagination* (1982), Alain
Corbin argued that over the eighteenth and nineteenth
centuries a fundamental shift in attitudes regarding waste,
privacy and hygiene occurred that impacted all aspects of
life in France.[188] Increasingly, according to Corbin, it was
smell – from the pleasure of a fine scent to the visceral
disgust induced by a foul one – that defined the parame-
ters of the social world.

At the behest of Emperor Napoleon III, the topography of Paris had been transformed over the course of the
nineteenth century by a programme of bold renovations.
These were led by Baron Georges-Eugène Haussmann,
an ambitious urban planner who replaced the warren of
streets and alleys that had characterised the medieval city
with a system of wide, straight avenues flanked by impressive buildings and interspersed with large public parks.
Renovations continued below ground, with the installation of *les égouts*, a vast network of sewers, and with the
laying of pipes to bring clean water into the city from
newly constructed reservoirs. Finally completed in 1870,

Haussmann's programme of works, which ripped up half the city, destroyed whole neighbourhoods and divided opinion then as now, establishing Paris as an exemplar of urban modernity. The 'Great Stink', which descended ten years later, threatened to unravel the city's image as a clean, safe and rationally structured space. More than merely a blow to civic pride, the faecal stench of the streets compromised people's sense of themselves. It was, as Barnes has it, 'no minor offense; it undermined a basic element of the invisible foundation on which modern civilization rested: the protection of the human senses from all that was base, vulgar and suggestive of bodily function'.[189] This psychological distance was made possible by the sanitary topography illustrated by Figure 17: vast subterranean structures whose pivotal role in the daily life of the city could remain unseen and unacknowledged by polite society. Along similar lines, one response to London's 'Great Stink' saw the installation of the city's first public toilets. In an 1858 letter to the Commissioners of Sewers, George Jennings – who had debuted the flushing toilet at the Great Exhibition – insisted that he was the right person to lead the effort, writing that 'I think it only right to call attention to the efforts I have made to prevent the defilement of our thoroughfares and to remove those Plague spots that are offensive to the eye, and a reproach to the Metropolis'.[190] He proposed that public toilets be installed beneath the pavements, so that 'offensive' bodily functions could be hidden from view and set apart from life on the bustling streets above. In Paris and London, the fantasy of the modern city as an ordered, enlightened place, emblematic of the glittering achievements of the

Figure 17: Drawing by Albert Lloyd Tarter produced
in the early 1940s for an educational film on
disease and sanitation that was never made

present, could only be sustained by the disavowal of excremental processes.

The image that nineteenth-century society had of itself,
as a time of great social advancement, cultural refinement
and technological development, was built on a rejection of
the past as less clean than the present. When Haussmann
demolished medieval Paris, he aimed to destroy a way of
living and dealing with bodily waste that had become
incompatible with France's vision of itself as a civilised,
hygienic nation. This shift occurred, in part, because of
a hidden world slowly coming into view. New scientific
theories drew attention to the menagerie of microscopic

Figure 18: Horrified woman observing
water through a microscope, 1835

creatures living unseen among us and held them responsible for all manner of ills. Driven by the pioneering work
of figures like the German physician Robert Koch and the
British surgeon Joseph Lister, germ theory emerged in the
closing decades of the century as the dominant paradigm
for understanding disease. Popularising new conceptions

of the human body, grounded in the language of organism, bacillus and pathogen, emergent fields like bacteriology and virology emphasised the role played by microorganisms in the spread of illness. The discovery of this teeming microbial world was hailed as one of the defining scientific discoveries of the age, but it also proved profoundly unsettling. As a cartoon of a woman horrified by the creatures that she can now see cavorting in her drinking water suggests, knowledge of this hidden universe could provoke paranoia as easily as wonder (Figure 18). One response to this sense of unease was the increasing emphasis placed on cleanliness as both a moral virtue and an essential aspect of modern society, one that must be upheld at all costs. This sanitary revolution transformed not only sewer systems, then, but also the way people conceptualised dirt and disease.

Filth, as the enemy of human progress, must be conquered. In an advertisement for household disinfectant 'Anios' from 1910, 'le microbe' is declared 'l'ennemi' – winged and crawling beasts like 'cholera', 'mildew' and 'typhus' – that can be vanquished with the liberal application of disinfectant (Figure 19). The war on dirt was not, however, confined solely to battling with devilish microbes and ghoulish bacteria. Instead, the lower classes, those deemed feckless, criminal or otherwise beyond the pale, were characterised as somehow inherently dirty, responsible for polluting the urban environment and spreading disease. As bodily waste was removed from the public life of cities and excretion became a private matter, uncleanliness was increasingly viewed as a sign of degeneracy and a threat to the social order. In the late 1930s the German sociologist

Figure 19: Advertisement for Anios disinfectant
by G. de Trye-Maison, c.1910

Norbert Elias – whose writing on table manners was
touched on in an earlier chapter – attempted to understand
how seemingly objective forces, like the development of
sanitary technology, came to be so wrapped up with far
messier notions like moral purity and civic virtue.

His two-volume magnum opus *The Civilizing Process*
(1939) argued that historical development must be understood as a mechanism in which political change, economic
transformation and religious upheaval are always accompanied by revolutions in the social attitudes and mentality
of ordinary people. According to Elias, beginning in the
late medieval period, the trend in Europe was towards
ever greater feats of self-restraint, moderation that was
exercised in relation to everything from interpersonal
violence and sexual behaviour to table manners and conversation. Seeking to understand how significant shifts in
behaviour came about, he looked to the conduct manuals
that began to shape court etiquette from the end of the
fourteenth century onwards. One of these, *De Civilitate
Morum Puerilium* (*On Civility in Children*), written by the
Dutch philosopher Erasmus in 1530, includes lengthy pas-
sages on the correct way to dispose of snot, how to avoid
farting in polite company and the inadvisability of exposing in public those 'parts to which Nature has attached
modesty'. At the table, he advises against wiping greasy
fingers on clothes and warns that 'it shows little elegance
to remove chewed food from the mouth' and to return it
to the communal dish.[191] We might find it surprising to
find that the sons of lords and ladies had to be told not to
wipe their noses on tablecloths, but that is precisely the
point Elias is trying to make. Behaviours that appear to us

as natural and innate began life as a code of conduct that had to be studied and practised. Slowly these new ways of being were internalised, so that 'in keeping with the transformation of society, of interpersonal relationships, the affect-economy of the individual is also reconstructed'.[192] For Elias, what characterised the 'civilising process' above all else were rising levels of shame and disgust, especially regarding the body's waste. The proper management of those products – urine, wind, saliva and excrement – marked children from adults, the upper classes from the lower, and civilised from savage.

Higher standards of cleanliness and embarrassment around the exercise of bodily functions also help to differentiate past from present. Writing in the twentieth century, Elias identified the long-ago Middle Ages with filth and measured modernity in terms of its distance from that dirty old time. A similar impulse to distance the present from the past was at work at the International Health Exhibition (IHE) of 1884, a descendent of the Great Exhibition that featured a wide range of edifying and entertaining attractions intended to showcase developments in the science of health. Visitors to the IHE could attend nightly lectures on public health, dine in a vegetarian restaurant or sample tofu at the Japanese pavilion, observe live experiments at the Hygienic Laboratory or peruse stands displaying the latest sanitary innovations. By far the most popular attraction was, however, the Old London Street (Figure 20). A replica of medieval London, before the Great Plague of 1665 and the Great Fire of 1666, the street was composed of twenty-five buildings, painstakingly reproduced using historic drawings. Peopled by costumed performers and street

Figure 20: Two views of the Old London Street featured
as part of the International Health Exhibition in London.
Reproduction of a woodcut by A. Beresford Pite, 1884

vendors selling refreshments and souvenirs, the street was
entertaining, but it was designed with an edifying message
in mind. Stepping from the overcrowded and badly ven-
tilated environs of Old London into the airy spaces of the
main exhibition, visitors were encouraged to reflect on
how much progress had been made to the general standard of living. Encouraged to reflect on the sanitary follies
of their forebears, Victorian exhibition-goers defined their
present – characterised by hygiene, order and, above all,
the efficient disposal of bodily waste – against the disease,
disorder and filth of the past.

European identity was, however, forged not only in
relation to its own history; it also relied on the positioning of non-white, colonised peoples as being somehow

out of time with the Western world. This thinking was sustained by a particular understanding of historical progress, which the theorist Johannes Fabian has termed the 'temporal slope' of evolutionary thought.[193] Through a 'denial of coevalness' – a refusal to accept that we exist in the same timeframe as other people – Fabian argues that Victorian ethnographers positioned other cultures in the past. This meant that the 'civilised' observer was always imagined as occupying the 'present', characterised by industry, modernity, science, in relation to the 'past' of the delayed, 'primitive' subject defined by superstition, religion and myth. Nowhere was this temporal lag more apparent, for early anthropologists, than in attitudes and practices relating to excrement. The notion that all so-called 'savage' societies lacked excretory taboos became, according to historian Alison Moore, a fashionable topic towards the end of the nineteenth century.[194] Sometimes from direct observation, but far more often through second- or third-hand accounts, anthropologists noted the lack of shame attached to bodily functions and the use of waste in ceremonies. In her ground-breaking work of cultural theory *Purity and Danger: An Analysis of Concepts of Pollution and Taboo* (1966), Mary Douglas noted how Victorian anthropologists tended to characterise 'primitive' attitudes towards bodily functions as 'autoplastic' – meaning that they revelled in the body and its products – in contrast to the 'alloplastic' 'civilised' man who objects to faecal matter as filth.[195] Most disquieting were instances in which, usually as part of a religious ritual or to mark a rite of passage, human excreta was eaten. Reports of coprophagia confirmed, for those invested in the imperial project, the dangerous backwardness of colonised peoples and the pressing need to bring civilisation to the far-flung corners of the globe.

As exemplified by the vast subterranean sewers of cities like Paris and London, the advance of Western society relied on an urban sanitary order that controlled excretory procedures, removing them from view and allowing the enlightened citizen to distance themselves from such squalid concerns. The imaginary of the present, not only what the modern world looked and smelled like, but also its sense of itself as technologically advanced and socially refined, was sustained by the disavowal of certain bodily processes. Perhaps unsurprisingly, one outcome of this collective repression was that defecation became taboo, and thereby loaded with a potentially erotic charge. In the Marquis de Sade's notorious *120 Days of Sodom*, written in 1785 but not published until 1904, the story of four wealthy libertines who conduct an experiment in sexual gratification that involves a remote castle, kidnapped teenagers and escalating acts of violence, scenes of urine drinking and coprophagia abound. Waste can only become illicit and eroticised after it has been expelled from the scope of polite society. For Sigmund Freud, who was fascinated by tales of remote tribes and their unfamiliar beliefs, civilised culture was built on denying the pleasure of defecation. All children are, he noted, naturally drawn to the waste systems of the body, but part of healthy psycho-sexual evolution involved sublimating anal stimulation in favour of the genitals. This stage of development occurs around eighteen months, as the child discovers the erogenous zone of the anus and the pleasure that might be derived from controlling its movements. This discovery can place them in conflict with the adults in their life, who may be making their first attempts at toilet training. According to Freud, difficulties during this tricky transitional period – resistance to using the potty, withholding bowel movements, playing with faeces – can lead to neurosis in later life, as the individual finds ways to sublimate their anal pleasure into an excessive concern with cleanliness and order. In psychoanalytic terms, then, becoming an adult involves cultivating a sense of shame around toilet matters; shame about the body that eventually translates into appropriately restrained behaviour and social conformity.

Mapped onto culture and society more widely, this model of psychological development and self-restraint casts new light on the history of the gut. Suddenly, the great sanitary innovations of the nineteenth century seem to have less to do with public hygiene than with the fashioning of a civilised self. When Joseph Bazalgette completed London's sewer system in 1875 it was praised as the 'most extensive and wonderful work of modern times'; Baron Haussmann's rationalisation of Paris's underground network of pipes and tunnels was hailed as showcasing the latest engineering innovations. In both cases, effective sanitation was held up as proof of humanity's progress towards a cleaner environment, where infectious disease might be conquered, and bad smells forever expunged from the city streets. With this came an assured vision of the present – that is, the white, bourgeois, Western present – as somehow removed from the digestion and its products. Which relied, of course, on an understanding of the past as dirtier and its inhabitants as more enmeshed with the body's excremental processes than their Victorian counterparts.

This story could only be sustained, however, by ignoring the fact that modern Europeans came pretty late to the sanitation game. Excavations of the ancient city of Eshnunna – located north-east of Baghdad – exposed brick sewers and flushing latrines dating from around 2500 BC; residents of the Indus Valley in Pakistan enjoyed household toilets between 2600 and 2501 BC; and the Minoans used their advanced knowledge of hydraulics to install city-wide drainage systems by 1700 BC. As the American sanitary engineer Harold Farnsworth Gray put it in 1940, when 'the Englishmen of the last century finally did succeed in divorcing themselves from immediate juxtaposition with their own excrement and the stench thereof' they were only 'at long last catching up in part with the Minoans of nearly 4,000 years previously'.[196] That from the perspective of the ancient world nineteenth-century Europeans would have appeared a very backward people, only just beginning to grapple with the question of how to manage the waste of their growing cities, reveals something of the complicated and contradictory forces at play in the imaginary of the present.

In *Purity and Danger*, Mary Douglas famously described dirt as being simply 'matter out of place'. Despite provoking in us powerful 'cognitive discomfort and reactions of disgust', dirtiness is not an objective attribute, it depends on context and usually involves the contravening of a 'set of ordered relations'.[197] Think of a hair floating in a bowl of soup: the disgust it occasions arises not from something intrinsic to the hair itself – still attached to a head it might even provoke desire – but because it is somewhere that it should not be. Douglas's 'matter out of place' draws our attention to the contingency of emotions – revulsion, disgust, aversion – that can feel universal, but part of what this chapter has revealed is that dirt is also 'matter out of time'. As the vast sewers built under the streets of nineteenth-century London and Paris began to transform the place of bodily waste in the ecology of the city, moving it out of sight and out of smell, excrement came to be viewed as almost anachronistic. By associating faeces with the non-European and non-white, the colonised and the poor, it became possible to stake out a form of cultural superiority based on the 'correct' management of the products of the digestive system. As we have seen, this fantasy was always a fragile one, sustained by an overweening faith in the scientific authority and the obstinate disavowal of all the ways that waste continued to make meaning in the modern world. Delving deeper into the spaces between science and superstition, the next chapter explores the gut's prophetic powers.

8

Future

In 44 BC Julius Caesar was stabbed to death at a meeting in the Theatre of Pompey. The plot, which was orchestrated by two senators, Brutus and Cassius, triggered a civil war, and eventually led to the downfall of the Roman Republic. Before the assassination, Caesar, a populist leader who had pushed through a raft of controversial reforms and recently declared himself *dictator perpetuo*, was warned of the plot brewing against him. In William Shakespeare's *Julius Caesar* (1599) he is cautioned to 'beware the Ides of March', the seventy-fourth day of the Roman calendar, a date marked by the sacrifice of a sheep to the god Jupiter and the Feast of Anna Perenna, festivities that signalled the end of the new year celebrations. It was also, more ominously, known as a day for settling unpaid debts. In Shakespeare's play the warning is delivered by a nameless character identified only as the 'soothsayer'. Accounts from nearer the time, however, given by Cicero, Plutarch and Caesar's first biographer Suetonius, point to an Etruscan by the name of Spurinna, who was known to practise the art of haruspicy. This was a form of divination that used the sticky entrails of sacrificed animals to foretell the future. The haruspex would closely examine the liver,

Figure 21: The 'Liver of Piacenza', used in a
lecture delivered by William Osler in 1913

intestine and stomach for signs of providence or misfor-
tune. Bad omens might be observed in the discolouration
of certain organs or in the strange oozing of others, while
divine approval could be read from the plump fleshiness of
a healthy bowel. Caesar's fate was tangled up in the viscera
of several dead animals. Not only was Spurinna's warning
derived from the inspection of entrails, but further bad
fortune was cast when an ox sacrificed by Caesar himself
was found to be missing its heart. A bronze model of a
sheep's liver unearthed in an Italian field in the 1870s gives
some insight into the complexity of this ancient prophetic
tradition (Figure 21). The liver is divided up into sixteen
sections, reflecting the sixteen astrological houses of the
Etruscan cosmos, each inscribed with the name of one or
more deities. From the careful observation of this miniature heavens, the haruspex was able to establish which
gods were angry, which were favourable and what the

future might hold. The 'Liver of Piacenza', as it is sometimes called, is also one of the earliest known anatomical
models. In a series of lectures on the 'Evolution of Modern Medicine' delivered at Yale University in 1913, William
Osler, a Canadian physician best known as the co-founder
of Johns Hopkins Hospital who we met in Chapter Four
trying to buy Alexis St. Martin's stomach, described these
divinatory practices as a kind of dissection. Like the
skilled anatomist, the haruspex inspected the organs with
'unusual care', noting any marks, swellings and malforma-
tions, probing the sticky interior of the body in search of
knowledge about life and death.[198]

The difference between modern anatomy and the
soothsayers of old, of course, is that we no longer believe
that examining the viscera can help us to predict the fate
of powerful men or foretell the future of great nations. Yet
something of the body's prophetic potential remains with
us in the science of the gut. According to a recent study
carried out by the Harvard Medical School, our microbiome is a soothsayer capable not only of predicting if we
will develop conditions like diabetes or IBS in the future,
but also of forecasting our deaths.[199] Having sequenced
samples of gut bacteria taken from a wide range of participants, the researchers found that, in some cases, those
microbes may prove more effective predictors of life
course than genomes. Elsewhere, a Finnish study made
of thousands of stool samples donated in 2002 found that
those with a wealth of Enterobacteriaceae – the family of
infectious bacteria that features salmonella and E. coli –
were more likely to die within the next fifteen years.[200]
Tracking the lives of participants over this period, the

researchers found that even when lifestyle factors were taken into consideration, such as smoking or poor diet, the peculiar composition of an individual's microbiota did seem to exercise some influence on mortality. Our fate might even be sealed in the very early years of life, as the bacteria we acquire on our way out of the birth canal, from breastfeeding or even from the mouths of slobbery family pets, help to shape our immune system. The idea that the future course of your health and the possible date of your death might already be written in the body's bacteria, viruses and fungi is somewhat unnerving, but, as we've seen, the stomach has long been famed for its prognostic powers.

Modern understandings of the gut owe an unacknowledged debt to centuries of 'superstitious' thinking, magic and folklore, which have insisted on the organ's remarkable prophetic capacities. Haruspicy was also used by the Babylonians and the Ancient Greeks, where anthropomancy – the use of entrails from sacrificed men and women, often virgins – was employed to predict the outcome of major battles or chart the rise and fall of rulers. Less gruesome was the practice of gastromancy, taken up with enthusiasm in medieval Europe and in use well into the eighteenth century, which involved listening carefully to the gurgles and splutters of the stomach. Gastromancy was both a diagnostic tool used to identify illness in the body and a prognostic device that could divulge news of coming events in the world. It was advisable to treat the language of the belly with caution, however, as it was not always clear who or what was speaking. As Jan Purnis has uncovered, for most of history the practice of ventriloquism was associated not with the lips, but with the stomach. Derived from the Latin *venter* (belly) and *loqui* (speak), a ventriloquist was 'someone who speaks out of or from the belly, stomach or abdomen'.[201] Often treated as a religious authority, the ventriloquist was gifted with the ability to channel the voices of the dead through the stomach and foretell the future by interpreting its sounds. Throughout antiquity the gift of gastromancy commanded respect but, as we saw in a previous chapter, by the early modern period some had come to interpret the talkative gut as evidence of demonic possession. In 1656 the lexicographer Thomas Blount defined the ventriloquist as 'one that has an evil spirit speaking in his belly, or one that by use and practice can speake as it were out of his belly, not moving his lips',[202] and elsewhere the practice was increasingly interpreted as a kind of trickery, as the author of *A Perfect Discovery of Witches* (1661) complained:

This imposture of speaking in the Belly hath been often practised in these latter days [. . .] to draw many silly people to them, to stand wondering at them, so they by this imposture do make the people believe they are possessed by the devil, speaking within them, and tormenting them, and so do by that pretence move the people to charity.[203]

Only in the eighteenth century did the practice begin to resemble what we now think of as ventriloquism, as a deliberate performance intended to entertain rather than inform, and as one associated with the voice box rather than with the belly.

While some future seekers rooted around in the viscera, others have examined bodily waste for signs of what lay ahead. The practice of scatomancy, also known as copromancy or spatalamancy, transformed urine and faeces into remarkable divinatory tools. Throughout history and in cultures around the world, people have made careful observations as to the quantity, size and colour of human or animal excrement to read individual fortunes and glean glimpses of the world to come. In the late nineteenth century, John Gregory Bourke, a captain in the US Army stationed in the Territory of New Mexico, was inspired to investigate these prophetic customs after an odd evening spent in a local village. On 17 November 1881, Bourke was invited to attend a ceremonial dance held by the Nehue-Cue, a secret order of the Zuni people, a Native American tribe who trace their ancestry to farmers who settled in the valley of the Little Colorado River sometime in the last millennium BC. Bourke was an enthusiastic amateur ethnologist, who had already published several books documenting his travels in the Old West and encounters with Indigenous peoples, and jumped at the opportunity to observe the sacred rites. These turned out to be not quite what he expected. To begin with, the dancers were not attired in traditional costume; instead, several of them were wearing cast-off army uniforms and another appeared to be dressed as a priest. As he later recalled, 'The dancers suddenly wheeled into line, threw themselves on their knees before my table, and with extravagant beatings of breast began an outlandish but faithful mockery of a Mexican Catholic congregation at vespers.'[204] Next a feast was announced: women entered carrying dishes of tea, sugar and, to the captain's disgust, a cooking pot filled with urine, from which the dancers took turns drinking. Unsettled by this strange communion, Bourke was determined to know more and so began a decade of research that culminated in the publication of *Scatalogic Rites of All Nations* in 1891. Running to almost five hundred pages, the book covers topics as diverse as 'cow dung in religion', 'excrement gods of the Romans and Egyptians', 'urine in ceremonial ablutions' and the 'use of bladders in making excrement sausage'. This vast study of excrement also devotes a chapter to divination, omens and dreams that gathers examples collected by fellow ethnographers and anthropologists which seem to reveal the persistence of scatomancy across time and space. Alongside the Peruvian medicine men who tell fortunes using sheep dung, the reader learns that among the French peasantry it is considered good luck to dream of *merde*, that the Kamchatkans of Russia believe that wetting the bed portends the arrival of foreign guests and that sixth-century Europeans practised a form of augury that involved bird poo.

Though clearly fascinated by this wealth of feculent rites and excremental observances, Bourke was also keen to distance himself from such 'primitive' customs. The book begins by alluding to 'difficulties surrounding the elucidation of this topic' that arise, in part, because the 'rites and practices herein spoken of are to be found only in communities isolated from the world', but also because they are of such a shameful nature that 'even savages would shrink from revealing [them] unnecessarily to strangers'.[205] Positioning himself as an educated observer of cultures, religions and races other than his own, Bourke exemplified the 'denial of coevalness' described by Johannes Fabian in the previous chapter – by which colonised subjects are imagined as occupying a time other than that inhabited by white Europeans – by characterising the scatological as a phase of human development that the West has moved beyond. Though he cites the odd example from medieval Britain and occasionally notes the persistence of relevant superstitions among the poor of Europe, far more attention is paid to the 'repugnant' customs of 'primitive peoples in all parts of the world'. Illustrating what the historian Alison Moore has described as the 'universal relationship between excrement taboos and the civilising process', Bourke invokes the 'misuse' of waste as evidence of barbarism.[206] The strange mix of disgust and fascination that characterises *Scatalogic Rites of All Nations* may be bound up with the conditions of ethnographic study itself. As the literary historian Stephen Greenblatt has suggested, ethnography is a discipline whose 'enabling condition is the otherness of the object of study', so that negative emotions like revulsion and loathing sometimes come to play a role in the analysis; a role that is usually unacknowledged, but which becomes unavoidable in a lengthy study of excrement.[207] What Bourke's assured sense of superiority obscures, however, is that far from being the preserve of remote tribes and 'savages', scatology has informed the evolution of Western medicine and modern understandings of the body.

In 1526 the Renaissance polymath Heinrich Cornelius Agrippa published a searing critique of the state of scientific theory and practice. His *Declamation Attacking the Uncertainty and Vanity of the Sciences and the Arts* is particularly scathing of doctors and their diagnostic methods. Physicians are, he writes, the 'most exquisite judges of the ordure of men' who 'behold obscene and beastly sights with their noses and ears to hear and smell the belches, farts, stinking breaths, steams and stenches of the sick, with their lips and tongues to taste the black and loathsome potions, with their fingers to search the dung and excrements'. No wonder, he concludes, that most sensible people forbid them from 'coming to their table', so as not to risk having to eat from the same dishes as 'nasty physicians'.[208] An occultist and theologian who was deeply sceptical of all worldly reason, Agrippa mocked the idea that any knowledge might be derived from the gassy gurgles and putrid emanations of the corrupt body. Physicians were, he insisted, little more than common swindlers making 'divinations or prognostications' by rooting around in the excrement of their easily duped patients. Agrippa was not alone in his distaste and, as a cartoon from c.1890 reveals, the suspicion that doctors took a rather unhealthy interest in the eliminations of the sick was not only a concern of the sixteenth century (Figure 22). In the cartoon, a physician is so pleased with his patient's *'superbe'* stools, that the cheeky maid asks if *'Monsieur le docteur'* would like a fork? Blurring the line between scatology and coprophagy, the satirist points to something gleefully perverse in the medical profession's obsession with excrement. While studies like the *Scatalogic Rites of All Nations* insisted that the civilised world had moved beyond such 'primitive' passions, popular depictions of the healing arts suggested otherwise. Scatology was an essential diagnostic tool, as the close examination of urine and faeces could

Figure 22: *A Physician Examines a Patient's Stools*,
lithograph by Gustave Frison, c.1890

elicit important clues as to the health of the patient. When physicians probed excreta, it was not for glimpses of events to come, but insights into the fate of their patients.

For most of history, delving beyond the protective barrier of the skin was a very risky endeavour, so physicians made use of the body's evacuations to judge its health. Early modern doctors were particularly invested in the divinatory potential of bodily waste. In this period a great deal of attention was paid to excretion because, according to the historian Michael Schoenfeldt, the 'elaborate technology of digestion' was considered to dictate the overall health of the body. This was imagined to be a three-stage process that began in the stomach, but also involved the liver and a vast network of veins that transported nourishment to where it was needed. From this perspective, as Schoenfeldt writes, human anatomy was simply a 'giant stomach, a torus through which food passes' where everything came to depend on digestion.[209] The gut also played a pivotal role in humorism, the ancient system of medicine that we have encountered elsewhere in this book. Not only did the process of digestion help to generate yellow bile, black bile, phlegm and blood within the body, careful dietary management was also considered essential to maintaining humoral equilibrium. By eating certain foods and refraining from others it was possible to balance the humours – classified as hot or cold and wet or dry – a process that was fundamental to physical and mental health. If your humours were out of whack, perhaps an excess of black bile was making you melancholic or a lack of blood was sapping your energy, it would be obvious from your evacuations.

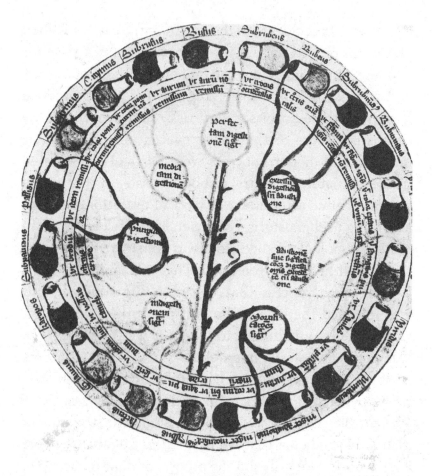

Figure 23: Urine chart dating from the
end of the fourteenth century

Uroscopy was one of the most common diagnostic techniques employed by medieval doctors of all kinds and involved the close sensory analysis of a patient's urine, using sight, smell and sometimes taste to determine the nature of their illness. Often the urine would be decanted into a glass flask – sometimes called a *matula* – which was designed to replicate the shape of the bladder to allow practitioners to observe the fluid in its natural habitat. Diagnosis would usually be made with reference to established diagrams that outlined possible colour and strata variations, their humoral significance, and disorders likely to arise from any potential imbalance (Figure 23). As historian Joseph Tate has uncovered, at their most ubiquitous, uroscopists were widely known as 'pisse-prophets', a derisive sobriquet that, nonetheless, goes some way to accounting for their sustained popularity as experts who were able to offer insight into the future health of their patients.[210] While the Liver of Piacenza transformed an organ of the body into a tool for unravelling the mysteries of the universe, urine charts unveiled the secrets of the body by decoding its waste. Though not quite as common as uroscopy, early modern physicians also practised scatology. In his *Praxis Medicinae* (1639), for instance, Gualtherus Bruele instructed the reader on how to diagnose disease by reading the shape, texture and smell of faeces. We learn that a 'hot distemper' usually evidenced a rather dried-out specimen, while in a 'cold distemper' it tends to 'waxeth thick and more tough'; that 'weakness of the brain' can be detected by the presence of half-digested matter; and that death is often foretold by the voiding of 'black excrement'.[211] Humoral medicine, with its faith in processes of letting and purging, hailed excrement as a kind of soothsayer, but it was not the only ancient system to do so. In Ayurveda, a tradition rooted in the Indian subcontinent that has been traced back as far as 6000 BC, similar attention is paid to the size, shape, smell and frequency of eliminations as valuable guides to mental and physical wellbeing. Ayurvedic medicine emphasises the need to manage the elemental forces in the body, known as *vāta*, *pitta* and *kapha*. Achieving balance or *sāmyatva* between the three results in health, while imbalance or *viṣamatva* results in disease. Waste management – of both *sharirika mala* (body wastes) and *dhatu mala* (metabolic wastes) – is thought be essential to maintaining bodily equilibrium, and stool analysis remains an essential tool for Ayurvedic doctors.

Scatology is having something of a renaissance, as a chorus of alternative practitioners, wellness gurus and innovative health start-ups encourage us to pay more attention to our poo. Writing for the *Guardian* in 2021, Hannah Marriott described how the 'rise of stool-gazing' has been fuelled by campaigns like The Gut Stuff – whose adverts, posted on bus stops around London, asked commuters whether 'smooth criminal, the smashed avo [or] the poonami' best described their evacuations – by apps like the Moxie Poop Scanner, which promises to help the user to correctly categorise their stools, and by companies like Seed, which has set up a database onto which it is possible to upload pictures of your poo that will help train artificial intelligence to distinguish between healthy and unhealthy excrement.[212] These commercial ventures are all based on appropriating and redeploying a clinical assessment tool called the Bristol Stool Chart. Developed in 1997 by doctors based at the Bristol Royal Infirmary, the scale organises human faeces into categories ranging from the hard lumps of Type 1, past the smooth healthfulness of Type 4 and on to the watery formlessness of Type 7. Now widely used in the diagnosis of gastrointestinal disorders, it has proved especially successful in the early detection of IBS and has been taken up as a tool for measuring the effectiveness of a range of ongoing medical treatments. The invention of the Bristol Stool Chart was not, however, the first time that modern medicine had attempted to schematise faeces. Published in 1921, René Goiffon's *Manuel de coprologie clinique* could be read as a precursor of sorts. In his coprology, the French doctor proposed that the quality of the patient's faeces had little to do with the food consumed and was rather determined by the peculiarities of the course it plotted down the oesophagus, into the stomach and through the intestinal coils. While his early modern forebears relied only on the information of their senses to map this journey, Goiffon was able to observe the action of bacteria at different stages in the digestive process and to identify the complex structural components of faeces using microscopic analysis.[213] Today the emergence of 'stool-gazing' as a popular pursuit speaks to a similar fascination with how waste might make our innermost workings visible and, in some new way, comprehensible.

Wonderful as such knowledge might be, it does not really compare to the nation-shaking prophecies and terrifying visions of the future that were once attributed to the gut. It is possible to read the history of medicine as a story of disenchantment in which the magical potential of the body – as something that might transform or transcend the conditions of the everyday – was slowly erased by the rise of scientific rationalism. On first inspection, this seems to have been the fate of the oracle stomach, imbued by the ancients with powers capable of bringing down civilisations and then reduced by the modern world to commenting on minor digestive upsets. Yet the relationship between time, medicine and the gut remained more complicated than this apparent demotion allows for. One area that this becomes apparent is in the long history of dissection and the story of the modern autopsy. In 1531 the translation and republication of Galen's *On Anatomical Procedures* reignited interest in the procedures and possibilities of dissection by making practical instruction widely available for the first time. This revival was also aided by a shift in the kind of bodies that it was deemed acceptable to use in the teaching of medicine. Because the use of human bodies was largely forbidden in Ancient Greece and Rome, classical works based their knowledge of anatomy on the observation of different animals. But by the time of the European Renaissance, anatomists were granted access to a ready supply of cadavers – usually executed criminals – and were able to accurately map the inner structures of the body for the first time.

In 1543 the Flemish physician Andreas Vesalius published his *De Humani Corporis Fabrica* (*On the Fabric of the Human Body*), seven books that described every part of the body from the bones and the muscles to the veins, the heart and the brain, in rich and vivid detail. This remarkable verisimilitude was achieved, in part, by the inclusion of woodcut illustrations, drawn from direct observation, depicting the anatomical body in different poses (Figure 24). Based on a series of lectures Vesalius delivered to medical students at the University of Padua, the book was a teaching aid that not only described the body, but also provided step-by-step instructions on how to locate, examine and dissect its organs. The first lesson that every

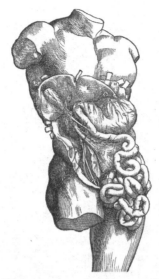

Figure 24: Illustration of a dissected torso from
De Humani Corporis Fabrica, Andreas Vesalius (1543)

budding anatomist had to learn concerned the order of dissection: how and where to begin. Some theoretical works recommended starting with the head and finishing at the feet, while others proposed that the body be sliced down the middle and opened like a book, but practical guides like *De Humani Corporis Fabrica* always started with the gut. As cultural historian Jonathan Sawday explains in *The Body Emblazoned: Dissection and the Human Body in Renaissance Culture* (1995), 'The anatomist dissected the body according to its rate of decay, commencing with the abdominal viscera, and then moving on to the thoracic cavity, the head, and, finally, the external members – the limbs.'[214] In other words, because the stomach, intestines and colon decayed quickly after death they had to be examined first. So, in the grand anatomical theatre of

Figure 25: Scale model of anatomy
theatre built at Padua in 1594

Padua, where onlookers gathered to watch learned men like Vesalius unveil the deepest mysteries of the body, it was the messy viscera of the digestive system that ultimately set the pace (Figure 25).

More than simply the anatomist's timekeeper, the stomach could also be called upon to reveal secrets about the last moments of its owner. The origins of what we would now recognise as the modern autopsy can be traced to the late eighteenth century and the gut played a pivotal role in the development of the discipline from the beginning. Fuelled by the publication of key texts like Giovanni Battista Morgagni's *The Seats and Causes of Diseases Investigated by Anatomy* (1761) – which was the first to correlate findings after death with the clinical picture in life – and later made possible by developments in microscopic technology, pathological anatomy emerged as the only scientific field dedicated to understanding death. By carefully examining organs and tissues for signs of disease, noting internal changes alongside external signs and symptoms, it became possible to retrace the steps that led the deceased to the mortuary table. For the post-mortem detective, the gut could provide several important clues. Most obviously, in cases of poisoning or overdose the stomach might directly reveal the cause of death; and even when no trace of the noxious substance remained, changes in the colour and appearance of the other digestive organs could provide evidence that something unpleasant had passed through the system. Beyond toxicology, the stomach has also been called upon to help determine the time of death. The classical method for estimating this involves measuring post-mortem changes to the body like algor mortis and rigor mortis (cooling and stiffening), maceration (tissue softening due to the presence of liquid), putrefaction, discoloration and examining the gastric contents. Though digestion varies from person to person, a meal is typically fully digested between four and six hours after eating, and this fact has often been used to help calculate the time of death. Paused at the moment of death and filled with the half-digested remnants of its last meal, by the twentieth century the stomach came to be imagined as a kind of gastronomic time capsule. Today most forensic pathologists are wary of relying too heavily on gastric analysis as a technique, especially as part of a criminal investigation, but there remains something appealing in the image of the gut as a record of the world buried deep in the abdominal cavity. In the autopsy room the ancient oracle stomach retains some of its potency as a seer of unseen events, revealing the secrets of the dead through its contents.

The term autopsy, derived from the Ancient Greek αὐτοψία, 'to see for oneself', conjures a form of extraordinary knowledge of not only the inner world of the body, but also the world – one peopled with friends and enemies – that the living body has moved through. The first surviving record of an autopsy is of the one that was carried out on the corpse of Julius Caesar. Following his assassination in the Theatre of Pompey, his corpse was gathered up by three slaves and returned home, where his personal physician Antistius took over. Having examined him and recorded twenty-three stab wounds, the doctor concluded that the fatal blow had entered just under the left shoulder blade, either striking the heart or nicking a major artery, causing death by massive internal bleeding. From the very beginning, then, there were surprising connections between the prophetic arts and the medical practice of autopsy: Spurinna the haruspex observed Caesar's future in the guts of sacrificed animals and then Antistius the physician retraced the events of the recent past in his corpse. Later, the arts of uroscopy and scatology continued to blur the line between diagnosis and divination, meaning that the history of medicine is, at least in part, also the history of magic, enchantment and prophecy. Modern research into the microbiome exemplifies this duality: cutting-edge science underpinned by a faith in the gut as the augury of the body, capable of predicting disease and forecasting future health. For all those willing to place their confidence in the prognosticative power of the belly, there have been others who have met its proclamations with suspicion. Whether founded on the possibility of demonic possession or on the dismissal of non-Western medical practices as somehow more 'primitive', this scepticism has often betrayed a discomfort with the idea that the organs associated with the grubby business of digestion might have some larger role to play in the life course of a person, let alone in the affairs of the state, the fate of great leaders and the course of nation-shaking events. But whether imbued with miraculous prophetic powers or trusted as a wise diagnostician, throughout history the gut has offered up alternative ways of interpreting and making sense of interior and exterior realities.

Today when we call on a 'gut feeling' or scrutinise our evacuations, we place a similar kind of faith in the belly as a soothsayer, sibyl, sage or oracle in possession of a special kind of wisdom. This chapter has examined the future as a key temporal dimension of the belly and we have already observed how the organ has been made to speak to the present, but the past also has a part to play in the cultural history of the gut. However, where the story told here has painted the gut in a fairly positive light, as something that we have used – in various ways – to make sense of the world's baffling complexities, the past has a different tale to tell.

9

Past

How far would you go to improve the health of your gut? If probiotics did not do the trick, would you consider a more radical solution? The Taymount Clinic, housed in a substantial red-brick building on the outskirts of Letchworth Garden City, is pioneering an approach to digestive wellbeing that goes far beyond the usual advice to eat more natural yogurt. Here it is possible to undertake a faecal microbiota transplant, a procedure that promises to restore gut flora and, in turn, reinvigorate the digestive system. Sometimes described as a human probiotic infusion or more colloquially as a poo transplant, the procedure involves taking intestinal bacteria from a healthy individual – the clinic boasts a bank of over two thousand stool samples for patients to choose from – and implanting them in the body of the patient via a rectal catheter or orally in the form of a capsule. This rather dubious-sounding procedure is based on the science of bacteriotherapy, an emerging understanding of the body that elevates the microbiome, the vast ecosystem of yeasts, fungi, viruses and protozoans that inhabit our gut, as a key player in the health of both body and mind. Faecal transplants have been shown to have some success in preventing the recurrence of gastric infections, especially those like *Clostridium difficile* (*C. diff*) which are exacerbated by the use of antibiotics, but the procedure's boldest advocates claim that it can be used to treat a far wider range of health problems. That the procedure is hailed, in some quarters, as a cure for everything from ulcerative colitis and Crohn's disease to multiple sclerosis, diabetes and even depression speaks to the power of the microbiome as a new way of imagining and organising knowledge about the body.

The discovery of this trillion-strong army of bacteria has been hailed as a significant advancement in contemporary medicine, but the assumption that digestive processes dictate our overall wellness has a far longer history in the field of alternative therapy. Today the Taymount Clinic is joined by a bevy of nutritionists, hypnotherapists, homeopaths, naturopaths, reflexologists and herbalists who have also chosen to set up shop in Letchworth. A proliferation that might be surprising elsewhere in the Hertfordshire countryside, but which is quite in keeping with the town's heritage. Built in 1904 by Ebenezer Howard, a social reformer who argued that new communities should be established between the city and the countryside, so that people could live apart from industry and have easy access to green, open spaces, Letchworth was intended to serve as a blueprint for the healthy cities of the future. An experiment in new ways of living, whose high street featured several health food shops, a vegetarian restaurant and a temperance pub called the Skittles Inn, from the beginning the town attracted a certain class of resident. The 'typical Garden citizen' was, according to a satirical sketch from 1920, 'clad in knickerbockers, and, of course, sandals, a vegetarian, and a member of the Theosophical Society, who kept tortoises which he polishes regularly with the best Lucca oil'.[215] Elsewhere, sharp-tongued writers like George Orwell lampooned the Letchworth 'type' as excruciatingly earnest bean-eaters with unpleasant 'wilting beards'.[216] Dismissed as cranks, these 'Garden citizens' spoke, nonetheless, to a broader set of attitudes regarding the pace of modernity, mechanisation and urbanisation.

Returning to the early decades of the twentieth century, this chapter explores how the workings of the gut came to be seen as simultaneously out of step with the rapid rhythm of the age and as representative of an older, more authentic state of being to which many aspired to return. This was a period of great upheaval and from the catastrophe of the Great War, when Europe had teetered on an economic and political precipice, came a shared desire to remake and reimagine the world. For many, this rejuvenation could only be enacted by going back to 'natural' ways of living, by getting back to the land and finding spiritual fulfilment there. This enthusiastic embrace of the great outdoors saw membership to cycling clubs soar, and rambling societies established, while organised nudism, nature-based mysticism and sunbathing flourished among the urban middle classes. The construction of towns like Letchworth was another expression of this turn to the rural as the site of fresh air and happiness. With this valorisation of the countryside came a demonisation of the city as a source of illness and unhappiness. As has already been touched upon in relation to the history of the working lunch, the rapid growth of cities – overcrowded and often polluted – led many to wonder whether the urban environment might not be conducive to the health of their residents. Focusing on the work of a surgeon who became notorious for his obsession with the ill effects of modern life on the digestive system, this chapter looks at how the gut came to be understood as an organ under threat from the forces of the present.

One result of the collective turn to nature was a renewed focus on the importance of raw fruits and vegetables to good health, with the result that the consumption of what were referred to as 'sun-fired' foods hit an all-time high during the interwar period. Advances in nutritional science revealed vegetables to be rich in essential vitamins and, as a result, governments in Britain and the United States launched dietary education plans that emphasised the body's need for fresh produce. Reformers began looking to vegetables to build a healthy and vital national body, and this was an important project because, as the draft had revealed, a substantial portion of the population were far from fighting fit. What else was standing in the way of the nation's health? Not only poverty, dangerous working conditions and poor healthcare, but also the rather hazier threat of 'over-civilisation'. In the bustling metropolis man lost some essential connection to the natural world: living in heated homes, travelling to work in comfortable omnibuses and eating canned foods, twentieth-century man was becoming a soft and ineffectual creature. Modern life, according to health reformers, caused nervousness, lethargy and depression, but worst of all was the damage it inflicted on the digestive system. Just as medical scientists today look for clues in the microbiome, so their forebears identified the gut as a barometer

Figure 26: Advertisement for Carter's Little Liver Pills, 1920

of bodily and emotional health. One common complaint, costiveness, was held to be especially dangerous to well-being and several high-profile health reformers dedicated themselves to ridding the population of this unfortunate affliction.

The early twentieth century was constipation's *belle époque*. Sales of purgatives, tonics, syrups and patent medicines like Carter's Little Liver Pills (Figure 26) went through the roof as ordinary people were encouraged to closely monitor the frequency and quality of their bowel movements. In Britain the campaign against costiveness was spearheaded by a man named William Arbuthnot Lane, who spent much of his medical career popularis-ing the idea that bodily health depended on the regular elimination of its waste. A highly regarded surgeon based at Guy's Hospital in London, he undertook pioneering work in the field of orthopaedics and neonatal care before becoming interested in the mysteries of the gastrointes-tinal tract. Born in Inverness in 1856, Lane hailed from a long line of medical men and his early childhood was spent with his military surgeon father in war zones across the British Empire. He returned to Scotland to complete his education before, having taken a particular interest in the study of anatomy, he enrolled as a student at Guy's Hospital. He remained there for his entire career, where he distinguished himself as an expert in numerous medical fields. In addition to mastering the rigors of ear, nose and throat surgery, Lane designed several surgical instruments – bone-holding forceps, a periosteal elevator and an osteotome, a bone-cutting device – which are still in use today. In theatre he championed the use of strict aseptic procedures that reduced the risk of infection for patients. Following in the footsteps of his father, during the First World War he joined the Royal Army Medical Corps and helped to found Queen Mary's Hospital, where reconstructive surgery on returning soldiers was first pioneered. For his wartime efforts and in recognition of his remarkable contribution to modern medicine, he was honoured with a knighthood. As these achievements reveal, by the end of the war Lane was one of the best-known and most respected medical men in the country, and yet by 1920 he found himself embroiled in a bitter dispute with the British Medical Association, which would eventually see him struck off the medical register, forced to give up his lucrative private practice and ostracised by the wider professional community.

This dramatic fall from grace was the consequence of Lane's long-standing, almost monomaniacal preoccupation with constipation. More than a discomfort or an embarrassment, costiveness was, he claimed, 'the cause of all chronic diseases of civilization' and the most serious public health crisis facing the developed world.[217] The danger came from the threat of autointoxication, a medical theory which held that half-digested matter sitting in the gut for too long would begin leaking toxins into the rest of the body. Chronic intestinal stasis occurred, according to Lane, when 'the passage of the contents of the intestinal canal is delayed sufficiently long to result in the production of an excess of toxic material, and in the absorption into the circulation of a greater quantity of poisonous products than the organs which convert and excrete them are able to deal with'.[218] These 'poisonous products' were the cause of a whole host of seemingly unrelated health problems: from stomach pain and sallow complexion to muscle pain, cold hands, baldness, excessive perspiration, a compromised immune system, rheumatoid arthritis and even cancer.

Less quantifiable, but just as worrisome, were the mental and emotional problems that could arise from irregular bowel movements. Writing in 1916, Frank Crane, an American physician and another proponent of the theory of autointoxication, observed that most 'domestic friction' – misbehaving children, fractious spouses and cantankerous relations – could be attributed to the malignant influence of unhealthy bacteria lurking in the gut. What the religiously minded might be tempted to label 'sin' was, he insisted, often due to 'what the physician calls stasis'.[219] Beyond family squabbles, incomplete digestion could, in extreme cases, cause full-blown insanity. One of the founding members of modern psychiatry, the German physician Emil Kraepelin, was particularly adamant that many diseases of the mind originated deep in the viscera.

Dementia praecox, what we would now define as schizo-phrenia, was caused by 'an endogenous process of chronic autointoxication which led to a self-poisoning of the body and, eventually, its brain'.[220] Interestingly, recent research has explored possible connections between the micro-biome, cognitive function and the symptoms associated with schizophrenia. Having compared stool samples from sufferers and non-sufferers, a team from the University of Wollongong in Australia found dramatic differences in their microbial make-up and they have posited that this may impact the neuronal functioning of the brain.[221] While today scientists tend to view the gut–mind rela-tionship as a fundamentally friendly one, sometimes set off course by miscommunication, doctors working at the beginning of the twentieth century took a much dimmer view of this alliance. The theory of autointoxication ele-vated digestion as the primary determinant of physical and mental wellbeing, but it also recast the bowel as the primary site of pollution in the body.

The human digestive tract was, by Lane's estimation, the biological equivalent of the household drainage system: with the stomach as the toilet bowl, the small intestine the drainpipe and the large bowel the septic tank. Just as no one would allow 'the discharge of offensive and poi-sonous gases' into the rest of the house and would instead 'have the cesspool cleared of the material stagnating in it', so too with the guts. Like the vast sewage networks lying underneath the streets of cities like London and Paris, the intestines are responsible for channelling and processing the putrid wastes of the body. Ill health can, therefore, usually be attributed to 'defective drainage'.[222]

As this imagery of toxicity, filth and corruption suggests, Lane had an almost adversarial attitude towards the gut; a myopic suspicion of the digestive system that came to govern his approach to treating patients and which would eventually lead to professional disgrace.

He was, however, not alone in his mistrust of the colon and throughout the early decades of the twentieth century many physicians and health reformers also warned of the dangers posed by autointoxication. John Harvey Kellogg, the nutritionist, diet reformer and anti-masturbation campaigner, was similarly invested in bowel health. Indeed, the invention for which he is now best remembered, the cornflake, was developed with the aim of introducing more fibre into the American diet and thus stave off the epidemic of intestinal stasis that threatened the well-being of the nation. Kellogg's thinking on constipation, that anything less than three bowel movements a day risked dangerous intestinal putrefaction, was shaped by the emerging germ theory of disease and the ascent of bacteriology as a discipline through the closing decades of the nineteenth century. As we saw in Chapter Four, this new medical paradigm impacted not only how external threats were imagined – the hordes of viruses, spores and prions that now appeared to be always threatening to break through our defences – but also how the internal world of the body was envisioned. From the beginning attention was directed to the presence of microbes in the digestive system and questions raised as to their possibly malignant influence on health. Physicians like Robert Bell, credited with having first coined the term 'autointoxication', and Charles Bouchard, a French doctor who was a well-known authority on the gut, warned that hostile microbes lurking in the intestines might slowly, through processes of fermentation and putrefaction, poison their host.[223]

Most influential in the field was Élie Metchnikoff, a Russian-born microbiologist, zoologist, the director of the Pasteur Institute in Paris and, perhaps most famously, an early adopter of live yogurt as an aid to digestive health. Having studied the diets of communities around the world who were known for their exceptional longevity, Metchnikoff found that most featured the daily consumption of some kind of soured milk preparation and from this correlation concluded that lactic acid must have some part to play in securing a long life. Almost as suspicious of the digestive system as Lane, the immunologist condemned the gut as an engine of senility in the body, where poisons generated by intestinal flora fuelled the ageing process. Only by flooding the system with the good microbes found in yogurt was it possible to slow this decay because, as research undertaken by the great French chemist Louis Pasteur had already demonstrated, lactic acid stalled bacterial growth.

Once publicised, this remarkable claim set off a yogurt-eating frenzy in Edwardian Britain. New products like 'Vitalait' and 'Lacto-bacilline' hit the shelves and the popular press hailed Metchnikoff as a new authority on health. Others were less kind, as in an article from 1916 that mocked him as 'the modern Ponce de León searching for the Fountain of Immortal Youth and finding it in the Milky Whey', but the idea that intestinal flora might have a hand in the ageing process lodged firmly in the cultural imagination.[224] Yogurt drinks like Yakult and

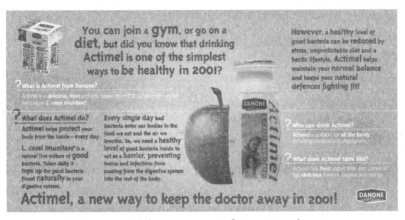

Figure 27: Advertisement for Actimel, 2001

Actimel contain *Lactobacillus paracasei*, one of a group of bacteria classified by scientists like Metchnikoff as 'probiotic': living microorganisms that, once consumed, 'top up [...] good bacteria'. As their pledge to prevent 'toxins and infections from passing from the digestive system into the rest of the body' (Figure 27) reveals, modern iterations of products like Vitalait make similar claims as to the health-giving, protective power of certain bacillus and the grave danger posed by a mismanaged gastric environment. Along with the shelves of fermented foods and flavoured kombucha found in health food shops around the world, probiotics hold out the promise that by carefully tending to the flora of the gut, it is possible to not only improve one's general health but also hold back the tide of ageing.

The theory of autointoxication advanced by health reformers like Lane and Kellogg engaged with not only the question of individual lifespan, but also more complex understandings of how the human species had evolved and adapted across vast stretches of time. Since the publication of Charles Darwin's *On the Origin of Species* in 1859, the theory of evolution had become the dominant paradigm not only for interpreting the natural world and human anatomy, but also for charting the development of culture. Outside of biology, evolutionary theory plotted the world's cultures and their peoples, past and present, along a scale from primitive to civilised. What this allowed, for white Europeans who inevitably identified Western society as being the most advanced, was the possibility of observing different stages on the evolutionary ladder at one time. For proponents of this kind of cultural evolutionism, life in an African tribe – in terms of social organisation, religious practice and so on – was simply less 'evolved' than that found in a bustling city like London. This was, for the most part, a cause for celebration, evidence of the natural superiority of the white race, justification for its colonial ambitions and a compelling argument for expanding the reach of the imperialist project. However, one notable exception in this self-congratulatory narrative of ever-onward and ever-upward progress was the stomach.

Civilisation was, most popular health advocates agreed, the enemy of good digestion. This was due, in part, to the pace of urban life, in which the stresses of the office, coupled with long commutes and irregular mealtimes, colluded to induce dyspepsia, constipation and a whole host of gastric turmoil in the weary urban populace. As we have already observed in relation to the history of the working lunch, life in the modern city was widely characterised as antagonistic to the gut's wellbeing. For Lane, however, the problem was more fundamental than simply a question of one's immediate environment. The trouble with the digestive system, he contended, was that it had not evolved at the same pace as the rest of the human body. The colon was, he insisted, essentially useless. Once needed to regulate elimination during fight or flight situations, now that much of humanity had developed beyond a minute-to-minute battle for survival, the organ no longer had a function. Worse than simply useless, it was also hazardous. Having originated in our four-footed ancestors, the colon had failed to adapt sufficiently when we evolved into *Homo erectus*, with the result that unnatural kinks had developed in the organ where toxic material lodged and festered. Walking upright slowed the transit of food through the system, meaning that waste material became lodged in the bands of the large intestine, setting the flora of the gut off balance and seeping poison into the rest of the body. Very much of the same mind, Kellogg condemned the colon as a 'poor cripple, maimed, misshapen, overstretched in parts, contracted in other parts, prolapsed, adherent, "kinked", infected, paralyzed, inefficient, incompetent. It is the worst abused and the most variously damaged of any organ of the body.'[225] Only humans and 'house dogs' who have their wild instincts tamed suffer the indignity of constipation, an ailment that did not – by Kellogg's estimate – exist elsewhere in nature.

According to proponents of autointoxication, not only was constipation a uniquely human problem, it was also far more prevalent in some societies than others. Lane attributed this disparity as due, in large part, to the invention of the toilet. So-called 'advanced' nations like Britain were much more prone to constipation than apparently 'savage' peoples, because while the former insisted on lounging on the loo the latter adopted a squatting posture that pushed the intestines up and supported them with pressure from the thighs, easing the passage of waste material and preventing backlogs. In addition to the pernicious practice of sitting to empty one's bowels, constipation was further exacerbated in the Western world by soft mattresses, overstuffed armchairs and corsets that weakened posture, by omnibus and train travel that upset the safe passage of food through the gut, by excessive brain work that directed energy away from the bowels, and by the adoption of increasingly sedentary lifestyles that sapped vitality, making digestive health impossible. The gastrointestinal tract was an organ dangerously out of time with the modern world, a prehistoric survivor trapped in a civilised body.

One way to reconcile the needs of the primitive stomach with the demands of modern life was to simply eat more slowly. This was the advice issued by an American named Horace Fletcher who, having been turned down for a life insurance policy on the grounds that he was overweight, examined his relationship to food and found it in need of reform. In addition to choosing what to consume with greater care, he also found that eating more slowly helped ease digestion. Eventually gaining the nickname 'The Great Masticator', throughout the early decades of the twentieth century Fletcher argued that every mouthful of food should be chewed a hundred times until it resembled a very fine paste, before being swallowed. Warning that 'nature castigates those who don't masticate', he urged his followers not to eat when anxious and never to eat on the move, precautions that aimed to bring the modern diner into line with the ancient rhythms of the stomach.

By contrast, Lane focused less on *how* to eat and more on *what* to eat. Encouraging his patients to adopt a diet filled with bran, legumes, fruits and vegetables, he emphasised the need to eat as naturally as possible. Canned, frozen or otherwise processed foods should be avoided on the grounds that these caused the bowel to become sluggish, while meat was prohibited on the assurance that in less 'civilised', predominantly vegetable-eating cultures the curse of constipation was almost unheard of. Members of the Vegetarian Society, who we have already observed trying to market fleshless lunches to Victorian office workers, advocated for a meat-free diet along similar lines. Calling on the language of comparative anatomy and evolutionary progress, vegetarian campaigners pressed recalcitrant carnivores to consider the physical differences, usually in terms of teeth and the digestive system, between predators and herbivores. An 1895 contribution to the *Vegetarian Review* urged those sceptical of the benefits of meatless dining to consider the human mouth, in which teeth for grinding 'grains and other hard vegetable substances' predominate, and where the biting teeth are 'eminently well adapted for nibbling fruits' but not equipped to deal with the excessive demands of chewy flesh.[226] Writing in the same vein in his *Manifesto of Vegetarianism* (1911), C. P. Newcombe, the editor of the Vegetarian Society magazine and prominent temperance campaigner, called on the opinions of 'professors of comparative anatomy' to argue that man was 'originally' vegetarian. 'Man', Newcombe wrote, is 'made an intelligent being of the first rank – he has not to catch his prey like a wolf; he has no claws with which to hold, nor fangs with which to tear' and as such, his digestive system is not robust enough to process the flesh of other animals. What is more, because of the relative length of their intestinal tract, when humans do insist on eating meat it remains 'detained too long in the system', slowly putrefying and eventually becoming a source of disease.[227] Here the consumption of flesh signified, beyond any moral concern for the welfare of animals or spiritual anxiety for the fate of the soul, simply a betrayal of basic biology.

Eating out of time with the antediluvian stomach was, reformers insisted, the source of many of the debilitating health problems plaguing modern Britain. Diseases of civilisation, like dyspepsia, weak nerves and worst of all constipation, were the result of ignoring evolutionary imperatives and continuing to consume undigestible flesh. In 1925 Lane founded the New Health Society to improve the fitness of the nation and halt the spread of preventable chronic conditions. Based in London, it was devoted to the practice of social medicine, which advocated lifestyle and dietary change as sure routes to better wellbeing. It treated individuals suffering from digestive disorders, but its broader aim was to transform a degenerating society into a vital, vigorous community. Taking up the language of social Darwinism and eugenics, the New Health Society prescribed a regimen of outdoor exercise, sunbathing and high-fibre foods to its patients. If auto-intoxication was the result of over-civilisation, then the cure must lie in returning to simpler ways of living: pure foods, loose clothing and plenty of time spent in nature.

Alongside these changes to lifestyle, Lane also recommended that his patients take a daily dose of paraffin oil, which produced regular bowel movements but also had the unfortunate effect of causing excessive flatulence. He took it himself and forced it upon his whole household, which included a pet parrot and a small monkey.

Paraffin oil appears to have kept the monkey regular, but it did not work for everyone and in persistent cases of intestinal stasis, Lane proposed a more extreme solution. Between 1908 and 1925, he performed hundreds of colectomies, involving the partial or complete removal of the colon. He was not alone in viewing it as a dangerous and ultimately dispensable organ. Metchnikoff, for instance, made the remarkable suggestion that for the sake of their future health 'every child should have its large intestine and appendix surgically removed when two or three years of age'.[228] Having begun by experimenting with colon bypasses and finding them largely ineffective, Lane began recommending that the whole large intestine be taken out in cases of chronic constipation. He performed hundreds of these operations and quoted frequently from satisfied patients whose lives had been transformed by it, but there was a great deal of debate within the medical community as to the efficacy of the colectomy. Lane was too extraordinary a surgeon to be shrugged off as a crank, but before long reports of worryingly high mortality rates began to circulate and several damning exposés appeared in prominent medical journals. Elsewhere, in the United States where the theory of autointoxication had spawned a thriving consumer market for tonics, pills and other cure-alls, moves had already been made by the government to stem the tide of misinformation: in 1906 when the American Medical Association set up a Propaganda Department its first action was to denounce it as quackery.[229] In 1913, after Lane had been removing his patients' colons for five years, a panel of the Royal Society of Medicine was finally convened to interrogate the wisdom of resorting to surgery to combat chronic constipation. Working under the title 'A Discussion on Alimentary Toxaemia', speakers from surgery, medicine, anatomy and bacteriology spent nearly two months deliberating over the validity of auto-intoxication as a theory and the value of the colectomy as a practice. Interrogating the idea of intestinal kinks, one participant suggested, tongue in cheek, that a giraffe be dissected to determine whether it, as a similarly vertical animal, suffers from the same crimped intestines, while another challenged the idea that the colon was a useless organ and claimed that all Lane had demonstrated with his surgeries was that 'no colon was better than a diseased one'.[230] These questions opened up the very real possibility that a venerated surgeon had performed hundreds of pointless, dangerous procedures on his patients and the tide of opinion began to turn against him.

According to historian of medicine James C. Whorton, by the 'early 1920s, Lane had become a butt of ridicule in his profession', mocked for his obsession with regularity and no longer taken seriously as a surgeon.[231] Yet even after his public fall from grace, Lane remained convinced of the dangers that constipation posed to health. Despite committing his life to the profession, in 1933 he cut ties with the British Medical Association so that he could continue to see private patients and publish on the topic of autointoxication. While it might be tempting to dismiss him as a fanatic whose all-consuming obsession with the frequency of his family's bowel movements tarnished an otherwise glittering career in surgery, that would involve ignoring not only the countless patients who sought his advice, but also the other physicians, popular health reformers and prominent scientists who were similarly committed to the theory of autointoxication. A more complex interpretation of this history would point to the way that, through the early decades of the twentieth century, a particular narrative around the gut began to take shape which pictured it as a storehouse for dirt and danger in the body, a leaky vessel always threatening to spill its contents. Characterised as a vestige of the distant past lingering on in the present, the stomach and the colon in particular were viewed as dangerously out of time with the pace of modernity.

Though no one is advocating for the surgical removal of large parts of the digestive system today, the stomach continues to be imagined along remarkably similar lines. From warnings that modern life 'depletes our gut microbes' to the purported health benefits of the slow food movement, the gut remains somewhat at odds with the rhythms of present.[232] Like Eden after the Fall, the modern stomach has become, for many alternative health practitioners, a source of corruption in the body, responsible for causing everything from migraines to asthma, eczema, chronic fatigue syndrome and rheumatoid arthritis. According to proponents of the so-called 'leaky gut syndrome', for instance, the organ can wreak havoc around the body because its walls, under certain circumstances, can become permeable. The interior of the bowel is lined by a single layer of cells, known as the mucosal barrier, a border wall erected to prevent potentially harmful germs from passing into the bloodstream. Gastroenterologists have found that, in some cases, disorders like Crohn's or coeliac disease can threaten the integrity of this structure, inflaming the bowel in ways that might exacerbate digestive discomfort. However, advocates of 'leaky gut syndrome' go further than science currently permits to claim that undigested food particles and bacterial toxins can move through the wall, triggering the immune system, causing persistent inflammation around the body, and even contributing to the onset of autism. This pathological permeability can seemingly be resolved through dietary change, usually the elimination of sugar, gluten and lactose, accompanied by nutritional supplements and herbal remedies. For those who refute its existence, 'leaky gut syndrome' is little more than a marketing ploy, a way to sell books and expensive probiotics to gullible consumers, but for those living with bad gastric health it offers the comfort of diagnosis and the hope of finding relief from debilitating symptoms.

What this controversy reveals, beyond questions of empirical proof and medical ethics, is that the rudiments of autointoxication remain with us in some form or another. The image of the bowel as a storehouse of rotting matter or a cesspool of toxic waste speaks to a deep disquietude with the act of eating, an uneasiness with the boundary crossing of exterior to interior that it involves. The gut has become the locus for a host of present-day concerns ranging from the increasing incidence of chronic illness to worries about processed foods, from rising levels of depression to concerns over the West's growing 'sleep deficit', in a way that echoes the early twentieth century's over-fixation on constipation as a source of disease. This has, as we have seen, a distinctly temporal dimension: by identifying the modern world as the source of poor digestive health it became possible to imagine the gut as an organ that was better served in the past. Reformers like Lane and Kellogg went as far as to characterise the intestines as vestigial, as the remnants of a period of human evolution long since gone, but it is possible to observe time at work in the body in subtler ways. Part of what the present, past and future of the gut has revealed is how interwoven our bodies are with the world around us. For Michel Foucault, a French historian of ideas known for his theorisation of the way bodies are shaped by societal forces, time is a 'technology of power' that has been wielded to discipline and control populations. Think of the strictly timetabled school day or the tyranny of the office time card, ways of structuring time that – according to Foucault – also force the functions of the body into patterns or rhythms.[233] This symbolic exchange has, as the closing chapters of this book will explore, profound implications for the politics of class, race and gender.

POLITICS

On 6 January 2021 a mob of over two thousand Donald Trump supporters invaded the United States Capitol Building in Washington, D.C., in an attempt to prevent a joint session of Congress from formalising the presidency of the Democratic nominee Joe Biden. The rolling news coverage captured acts of vandalism and looting as rioters, many of them armed, stormed offices and eventually invaded the Senate chamber. There were reports of assaults on journalists and police officers, pipe bombs were later found at locations around the Capitol, many people were injured and five were killed. The aftermath of these shocking events saw Trump impeached for a second time, criminal charges levelled against roughly half of those involved in the riot, a special committee convened and public hearings on the incursion broadcast on national television. Widely held to represent an attack on democracy itself, this attempt to overturn the results of a fairly held election through a combination of conspiracy theory and direct violence prompted a reckoning with the state of an increasingly divided nation. Among the hundreds of stories that circulated in the months that followed the insurrection, one concerning dietary requirements was reported on with surprising frequency. A man known as the 'QAnon Shaman', whose flamboyant appearance at the Capitol riot – shirtless, face painted, wearing a large horned headdress – had already garnered substantial media attention, was refusing to eat in jail. Asked in court to explain himself, the 'Shaman', otherwise known as Jacob Chansley, told the judge that he followed a diet that restricted him to only organic meat and vegetables. Consuming anything that was not organic would, he claimed, not only make him profoundly ill, but also be in direct violation of his shamanic religious beliefs. The judge duly granted his request for organic meat and vegetables. In the outcry that followed many commentators argued that an organic menu served to one white conservative man, while the rest of the prison population is not granted the privilege of choosing what to eat, perfectly exemplified the double standard that had emboldened a rabble of mainly white conservative men to run amok in the first place.

It is to the muddle of questions raised by the health-conscious conspiracy theorist that the next three chapters of this book turn. Firstly, there is something in the way that amid a major national conversation regarding the future of democracy attention kept being drawn back to what one man would and would not eat which points to the unexpected presence of dietary preferences and by extension the gut in the nation state. Secondly, that much of the ensuing media outrage centred on organic food as a marker of social and economic entitlement reveals how politically loaded the question of diet is. And thirdly, the image of a bare-chested man dressed somewhere between Viking and cowboy eating only pesticide-free produce invites us to consider how gender ideologies might be tangled up with the work of digestion. *Rumbles* has already made the case for the body and its functions as social and cultural concerns, but its closing chapters will demonstrate that the gut is also a profoundly political organ. One that has played an outsized role in the public sphere, whose functions have shaped world events in unexpected ways and dramatically impacted conceptions of race, gender and nationhood.

10

Body Politic

Automatons were all the rage in eighteenth-century France. Mechanical contrivances, usually modelling humans or animals, that were constructed and programmed in such a way so as to appear to move independently, enthralled the national imagination. Elegant mechanical figures danced at society balls, the 'Lady Musician' played for audiences around the country and *La Charmante Catin*, the 'charming doll', even made it to the royal court. Most famous of all was the *Canard Digérateur*, built by Jacques de Vaucanson in 1738. Vaucanson was among the most famous inventors in Europe and the *Canard Digérateur* – the Digesting Duck – was his crowning achievement. An intricate clockwork assemblage of over four hundred moving parts, including a smooth brass beak, webbed feet and fully boned wings that could flap, the machine was modelled on the close observation of the anatomy and movement of ducks in the wild. Rendered in gold-plated copper, it could quack, muddle water with its beak and rise up on its legs. Most remarkable of all, this metal mallard also appeared to eat, digest and defecate like a living animal. Promoting his marvellous creation, Vaucanson described how it was possible to observe 'all the Actions of a Duck that swallows greedily, and doubles the Swiftness in this Motion of its Neck and Throat or Gullet to drive the Food into its Stomach', before 'discharging it digested by the usual Passage'.[234] More than simply a demonstration of technical ingenuity, the defecating robot was built to settle a long-running scientific argument, touched on in Chapter Four, over whether digestion was a mechanical or a chemical process. Throughout the early decades of the eighteenth century, iatromechanists – for whom the grinding motion of the stomach was key – went head-to-head with iatrochemists, who claimed that food was assimilated with the help of acid and alkaline ferments within the bodily fluids. As we will see, this debate over our inner workings was closely linked to some of the period's key religious, scholarly and political controversies. A prominent member of the Académie des Sciences, Vaucanson was well placed to intervene in this dispute and set out to explore the nature of digestion. According to historian of science Jessica Riskin, he viewed his automatons not merely as entertainments but as 'philosophical experiments, attempts to discern which aspects of living creatures could be reproduced in machinery' and 'what such reproductions might reveal about their natural subjects'.[235] Peering through the duck's metal feathers, it was possible to glimpse the working stomach housed within and glean some knowledge of the animal's physiology in its simulacra.

When it was displayed at the Hotel de Longueville in Paris, the *Canard Digérateur* caused such a sensation that the philosopher Voltaire quipped, 'without Vaucanson's duck, you would have nothing to remind you of the glory of France'. Even though it was exhibited alongside two equally impressive automatons, a flautist, and a pipe player whose musical virtuosity might have made a more fitting tribute to national 'glory', it was the defecating bird that came to be hailed as a symbol of modern, enlightened France. Even when it was later discovered to be a hoax – actual digestion did not occur, Vaucanson had simply secreted breadcrumbs, dyed green, to simulate bird poo – the duck remained bound up with the nation's sense of itself as scientific, rational and inventive. The popularity of the *Canard Digérateur* reflects the broader investments of a culture that, as historian Emma Spary has it, 'regarded the stomach as a somatic locus where digestive, moral, and even political upsets manifested themselves and through which appetite was expressed'.[236] From table manners in the home to the cultivation of civility in the royal court, from famous culinary writers who shaped 'good' taste to physicians who instructed eaters on the immorality of overindulgence, digestive matters consistently mediated between individual bodies and the public sphere. Investigating how, in eighteenth-century France and elsewhere, the gut came to play such an outsized role in defining the character of civic life, this chapter is interested in where the language of the stomach meets that of governance, politics and power.

The connection between bellies and the nation state has sometimes proved more than metaphorical. Fifty years after the *Canard Digérateur* made its debut among the aristocracy, revolution swept across the country, bringing with it social upheaval, the dissolution of the monarchy and violent political turmoil. Food lay at the heart of the revolution. Marie Antoinette may or may not have said 'Let them eat cake', but it was undeniable that during the reign of her husband, Louis XVI, many ordinary people went hungry. The Women's March on Versailles, one of the earliest events in the French Revolution, was organised to protest against the high price and scarcity of bread in Paris. In the years leading up to the Storming of the Bastille a combination of crop failures and government mismanagement brought the country to the brink of famine. There were rumours of a plot to starve the population and the king's attempt at solidarity by eating mixed-grain *maslin* rather than the all-white *manchet* did little to quell rising public anger. Lack of bread, Thomas Carlyle reflected in his three-volume *History of the French Revolution* (1837), had not only engendered 'agitation, contention, disarrangement' among the people, it had also imbued them with a 'Strength grounded on Hunger' that fuelled the uprising.[237] Tracing the links between empty bellies and political turmoil, this chapter also considers where the demands of the stomach have influenced the events of world history. Taking up the idea of the 'body politic' as both a metaphor that represents the body as the government in miniature and as a way of describing a collective of citizens, it is interested in the place of the gut in the national body.

As aristocrats lost their elegantly coiffed heads across the Channel, members of the British ruling class fretted at the violent potential of their own disenfranchised and underfed population. Coming in the wake of the American Revolution and fuelled by the controversial ideals of *liberté*, *égalité* and *fraternité*, the popular revolt in France threatened to undermine the political stability of its closest neighbours. In this climate of paranoia, theatres around England quietly dropped one of William Shakespeare's more incendiary plays from their programmes. Throughout the closing decade of the eighteenth century, *Coriolanus* – a tragedy based on the life of the Roman leader Caius Marcius Coriolanus – was staged on barely a handful of occasions. As it told the story of a deadly famine and the citizen uprisings that follow, the play was considered too inflammatory to be performed in such tumultuous times. Its opening scene, in which an angry crowd protests against the price of grain and accuses its leaders of hoarding crops while the people starve, bore a striking resemblance to recent events in France. In the play, the mob are intercepted by an ally of Coriolanus, Menenius Agrippa, who tries to calm their anger by reciting one of Aesop's fables. 'The Belly and Its Members' describes an inter-bodily conflict between the extremities and the stomach. As Agrippa tells it, the belly is accused by the other body parts of lazing around – 'only like a gulf it did remain/ I' th' midst o' th' body, idle and unactive/ Still cupboarding the viand, never bearing/ Like labor with the rest' – while the limbs worked to feed it.[238] In an act of rebellion they decide to starve the greedy gut, but soon find that the whole body suffers when the stomach is not nourished. Wielded as a piece of political propaganda, Agrippa casts Rome's rulers as the belly and its protesting citizens as the 'mutinous members' who have failed to appreciate the wisdom of centralised leadership. Instead, ordinary people ought to 'digest things rightly' and realise that every 'public benefit' proceeds from the distributive

work of the stomach.[239] In a climate of unrest and uprising, it was deemed unnecessarily incendiary to stage a
speech in which a political leader chided ordinary citizens
for questioning the rule of their superiors.

'The Belly and Its Members' is among one of the earliest examples of the body politic, a metaphorical strategy
that maps social orders, political structures and economic
models onto the human body. In the oldest sacred book
of Hinduism, the *Rigveda*, the caste system is delineated
through the body: the priesthood is the mouth, soldiers
the arms, shepherds the legs and peasants the feet; in
Ancient Greece Plato equated the state with the human
body and stressed the need for each part to perform its
function in service of the whole; early theologians like St
Paul Christianised the metaphor by imagining Christ as
the head of the body and the Church, until reformers in
the late medieval period reclaimed it for the monarchy.
Across time and in wildly different contexts, the metaphorical body has helped to shape political discourse. In
Coriolanus the physiological importance of the stomach
to the overall health of the body is called upon to naturalise the hierarchical make-up of the Roman world, but
the fable of the belly has been put to other uses. In the
Middle Ages texts by philosophers like John of Salisbury
and Nicole Oresme ordered the organs of the body in
terms of their physiological and symbolic nature.[240] The
head sits at the top, associated with the ruling monarch
and the site of reason, intelligence and the soul; next the
eyes that see and the ears that hear are associated with
lawmakers; the functions of the heart correspond to wise
men; the stomach to the farmers who feed the nation;

the hands that defend the body are alike to warriors or knights; journeying merchants are the legs; and finally, the labourers who support the body are its feet. By the late eighteenth century, the political organisation of the body tended to associate the lower classes with digestion and the guts and the monarchy with thinking and the head, so that it became possible to shift the original meaning of 'The Belly and Its Members'. Where Aesop's fable had used the gut to symbolise the social value of centralised authority, in post-revolutionary France the parable was reconfigured so that it came to represent the power of ordinary citizens. As medical historian Bertrand Marquer has uncovered, by recasting the stomach as a 'major actor of the social body' reformist thinkers placed the labour of the working classes at the centre of civic life.[241]

Whether marshalled to maintain the status quo or realise social change, metaphors of the body have long shaped the rhetoric of state and nationhood. Not only have political and economic structures been imagined in corporeal terms, but the body has also been read as a miniature polity itself. Most clearly expressed in the ordering of the organs, from the virtuous head down to the base intestines, the politics of the body are perhaps most obvious in the frequent characterisation of the gut as something of a troublemaker. In the natural hierarchy established by early medical authorities, the stomach was expected to labour, obediently and without complaint, to sustain the higher functions. Good health required the stomach to know its place, but it was widely held that the organ had something of an attitude problem. In his *Medicinal Dictionary* (1745), Robert James – physician and close friend of the great lexicographer Samuel Johnson – warned that even minor 'Spasms of the Lower Belly' could impair 'Force of the Memory and Brightness of the Genius'. Despite its lowly position in the corporeal pecking order, the gut needed to be treated with the upmost sensitivity lest it throws the 'whole of the nervous System into irregular Motions'.[242] Reflecting on this need for deference in *The Stomach and Its Difficulties* (1852), the English physician James Eyre noted that while it ought to be 'invaluable, as a Slave' more often the organ becomes 'a dangerous, because too powerful Despot!'[243] Later in the nineteenth century the food reformer N. Henessey described it as causing 'headache, toothache, sickness, biliousness, nausea, indisposition, and innumerable other aches and sensations' and concluded that the gut was 'implacably opposed to man's progress and comfort'.[244] Held responsible for an astonishingly wide range of disorders, from polluting the blood and infecting the other organs with acidic gastric juices to bringing on dangerous heart palpitations and precipitating nervous derangement, the stomach was often imagined as an agent of disorder and disruption in the body, either an unappeasable tyrant or an uppity servant who has forgotten their place. While previous chapters of this book have lighted upon the gut's vulnerability to demonic possession and its sensitivity to excessive passions, here we have it imagined as a kind of political agitator that threatens the harmonious organisation of the rest of the body.

The frequent characterisation of the stomach as volatile and unwilling to work for the common good reveals barely concealed resentment at its outsized influence on overall health. Undeterred by its inferior status, the messy viscera of the digestive system continued to wield power over even the most elevated parts, interrupting the important work of the brain and infecting the mind with dark moods. As Thomas Carlyle, who was plagued by dyspepsia for most of his adult life, observed: 'It is not the *pain* of those capricious organs; that were little; but the irresistible depression, the gloomy overclouding of the soul, which they inevitably engender, is truly frightful.'[245] The anarchic belly seeds lawlessness throughout the body and it is made even more troublesome by the powerful role it is called upon to perform to maintain the individual's sense of self. Digestion always involves a kind of mediation between the interior and the outside world. To eat is to bring something exterior, a foreign substance, into the deepest recesses of the body. The act of consumption, then, entails not only physiological danger – the risk of poisoning or the possibility of an allergic reaction – but also the kind of psychological jeopardy involved in incorporating something 'not me' into the bounded structure of the body. Eating represents a moment of profound vulnerability, that can shake the foundations of personal and collective identity. According to the anthropologist Mary Douglas, who we encountered in Chapter Seven, what we choose to eat or not to eat is governed by our conception of 'symbolic pollution'. In the same way as dirt is simply 'matter out of place', a substance that provokes disgust because it violates the agreed-upon social categories, so too are our food choices informed by culturally established boundaries between clean and unclean, safe and unsafe, self and other.[246]

These borders are sometimes drawn along national

lines. Recall George Cheyne's warning in *The English Malady* that the price to be paid for overweening mercantile ambitions – for having 'ransack'd all the Parts of the *Globe* to bring together its whole Stock of Materials for *Riot, Luxury,* and to provoke *Excess'* – was a rise in nervous disorders. One result of all that ransacking had been the introduction of 'foreign *Spices'* to British cooking. This had proved disastrous for the nation's health, because such 'Provocatives are contrived not only to rouze a fickly Appetite to receive the unnatural Load, but to render a natural good one incapable of knowing when it has had enough'. The English stomach, accustomed to much plainer fare, was – Cheyne insisted – ill prepared for '*Eastern* Pickles' and exotic 'sauces', and likely to suffer terribly when such alien flavours were imposed on it.[247] Published in 1733 as the British Empire battled with its European competitors to dominate the globe, this dietary advice clearly had as much to do with matters of national pride and national identity as it did with individual health. A country's cuisine is a political matter or, as the anthropologist Claude Lévi-Strauss put it, the 'cooking of any society is a kind of language [. . .] which says something about how that society feels about its relation to nation and culture'.[248] Communicating meaning far beyond the plate, food is part, alongside history, religion, culture and so on, of the collective imagination. This is important because it illustrates, once again, the centrality of consumption, digestion and defecation to how we make meaning about ourselves and the world around us.

Perceived as a kind of 'language' for the nation, culinary traditions can provoke anxiety as well as pride. Pasta, for instance, has lately generated a great deal of controversy. In the early 1980s the Italian Academy of Cuisine deposited a recipe for *ragù alla bolognese* at the Bologna Chamber of Commerce that purported to be the authentic and true method for cooking the famous meat sauce. Ingredients are limited to beef, pancetta, carrots, celery, onion, tomato sauce, butter, olive oil, milk and half a glass of red wine, and instructions specify cooking the sauce slowly for three to four hours. Despite this effort, home cooks around the world continue to use dried spaghetti instead of fresh egg tagliatelle and to make unnecessary additions ranging in severity from sun-dried tomatoes to dark chocolate to instant coffee granules. A social media backlash was prompted in 2017 when the television cook Mary Berry instructed her viewers to add thyme, white wine and double cream to their sauce, and the *New York Times* faced similar approbation when it published a recipe for 'White Bolognese'. That a bowl of pasta can generate so much anger suggests that there might be more to it than simply sofrito, meat and olive oil. According to the philosopher Roland Barthes, certain dishes can come to symbolise the nation itself, how it perceives itself and the image it wishes to impart to the rest of the world. In his essay 'Steak and Chips', Barthes describes the meal as not only 'part of the nation' but the 'alimentary sign of Frenchness'. In other words, the symbolic properties of steak – that it represents both 'elegance' and 'bull-like strength' – are qualities that help to define the character of the French as a people.[249] Moreover, the ferocity with which Italians have attempted to protect the 'true' Bolognese reveals something of the role played by food in helping not only to create national identity, but also to defend it from any outside influence that might dilute or corrupt its perceived authenticity. The act of eating helps to create the national body, through shared traditions and common mythos, but it also threatens to unmake it.

What we eat and how we eat it can help distinguish insider from outsider, familiar from foreign, friend from foe. Nowhere is this more apparent than during wartime, when the national body finds itself vulnerable and assailed by enemies. During the Napoleonic Wars – conflicts spanning a twelve-year period between 1803 and 1815, throughout which an ascendant French Empire clashed with other European powers – recruiting posters in England were often illustrated with an engraving by William Hogarth (Figure 28). The etching depicts a scene at the port of Calais, as a weighty side of beef is carried to an English tavern in the harbour while a ragged cast of hungry-looking characters look on. Though the target of Hogarth's satirical eye was clearly Catholicism – hag-like fisherwomen adorned with crucifixes, the greedy friar pawing at the meat and in the background a group of superstitious locals kneeling before the cross in the grubby streets – when the print was reproduced as a piece of military propaganda, more attention was paid to its negative depiction of the nation's eating habits. Contrasting with the hearty joint of meat destined for English bellies, the French soldiers must make do with thin *soupe-maigre*, while an exiled Jacobite slumped in the foreground has been brought close to death by a meagre meal of raw onion and bread. The message to new recruits was clear: fight like the strong beef-raised Englishman you are or risk defeat at

Figure 28: William Hogarth, O *the Roast Beef of
Old England (The Gate of Calais)*, 1749

the hands of the weak effeminate soup-swilling French.
Its title, O *the Roast Beef of Old England*, was taken from a
well-known patriotic song originally performed as part of
Henry Fielding's play *The Grub Street Opera* (1731), the first
two verses of which go:

> *When mighty Roast Beef was the Englishman's Food,*
> *It ennobled our veins and enriched our blood.*
> *Our soldiers were brave and our courtiers were good*
> *Oh! the Roast Beef of old England,*
> *And old English Roast Beef!*

But since we have learnt from all-vapouring France
To eat their ragouts as well as to dance,
We're fed up with nothing but vain complaisance
Oh! the Roast Beef of old England,
And old English Roast Beef!

Nostalgic for an imaginary time passed when 'fathers of old were robust, stout, and strong' and the world lived in fear at the might of our 'Fleets', the song worries that the British have been reduced to a 'sneaking poor race, half-begotten and tame' and lays the blame squarely with enfeebling foreign cuisine, namely 'ragout'. Repurposed in aid of the war effort, the threat of French invasion was recast in digestive terms, as an attack on the nation's stomach as well as its defences, institutions and wealth.

In *Beef and Liberty: Roast Beef, John Bull and the English Nation* (2003), Ben Rogers has argued that during the eighteenth-century roast beef became emblematic of an increasingly confident English nation, set in opposition to the refined dishes of continental Europe.[250] The figure of 'John Bull', who started life as a satirical character and wound up as a much-beloved symbol of Britishness, was key to this new-found confidence. Usually depicted as red-cheeked and rather stout, wearing a tailcoat and some-times a Union Jack waistcoat, John Bull came to embody all the virtues – plain speaking, common sense, stubborn determination – that the British were, or perhaps would have liked to have been, associated with. As a personifi-cation of the nation state, he was dragged into political arguments over the decisions made in Parliament or fears over the declining fate of the country. Often, his stomach,

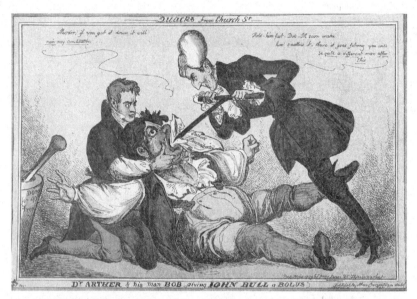

Figure 29: A large John Bull being held down and
force-fed by Peel and Wellington, representing the
idea of the Catholic emancipation as a breach of the
constitution. Coloured etching by W. Heath, 1829

replete with rich digestive metaphors, was invoked. Two
satirical cartoons, the first from 1829 and the second from
1913 (Figures 29 and 30), depict the force-feeding of John
Bull. In the first he is held down by Robert Peel, Home
Secretary, while the Duke of Wellington, naval hero and
then Prime Minister, shoves the Roman Catholic Relief
Act down his throat. In the second he appears strapped to
a chair as a line of figures representing everything from
'votes for women' and 'socialism' to 'home rule' and 'con-
scription', jostle to pour their issues into his medically
administered feeding tube. Produced in different contexts
and concerned with different political questions – the

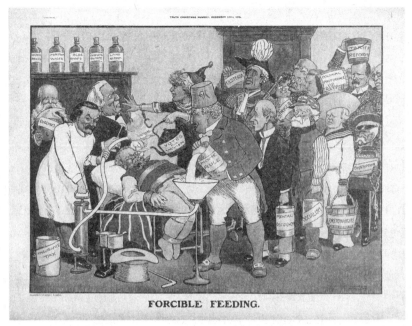

FORCIBLE FEEDING.

Figure 30: John Bull being force-fed via a
stomach pump. Stanger Pritchard, 1913

constitutional basis for Catholic emancipation in the first
and in the second, the pile of problems facing Britain in
the run-up to the First World War – the two cartoons both
show the violently overstuffed John Bull as a personifi-
cation of the national body placed under terrible strain,
made to swallow what is disgusting, what is intolerable.

As pieces of satire, these images were clearly intended
to induce the viewer to feel outrage or to act regarding the
issues of the day. It is significant that in each the prov-
ocation arises from imagined scenes of force-feeding, a
process long associated with the control and violent coercion of individual bodies by various institutions of power.

From enslaved Africans whose efforts at self-starvation were thwarted by the employment of a specially designed vice that wrenched their mouths open, to prisoners compelled to eat so they could serve their full sentences and asylum inmates who had tubes shoved down their throats to funnel nutrients directly into the stomach, force-feeding has been used to thwart dissent and assert the authority of the strong over the weak. We will return to the subject of force-feeding in the final chapter, but here it is enough to note how the state use of force-feeding to control bodies under its jurisdiction reveals something of the way that dominant ideologies have been articulated through the gut. The image of John Bull – representative of the national body – having unpalatable ideas, acts and demands shoved down his throat speaks potently of alimentation as a means of representing and expressing political debate or discord.

The violation of John Bull also reveals something of the violence at the heart of the digestive process itself. Eating can represent, as we have already seen, a moment of vulnerability in which the boundary of the interior self is breached by the exterior world. At the same time, however, if the eating of food is imagined as involving the absorption and domestication of foreign substances, then it could also easily serve as a metaphor for strong governance. According to historian Joyce L. Huff, the 'construction of alimentation as the incorporation of outside elements into the body's very structure meant that alimentary processes were easily expressed through imperialist metaphors. Indeed, the language of subjugation and conquest is everywhere in [. . .] discourses on eating

[. . .] food was incorporated into the overall structure of the body, just as colonies became part of the empire.'[251] In other words, digestion has often been envisioned in terms that pit the might of the gut against the incursion of what is consumed in ways that mirror the ideology of empire building. This version of digestion as always involving subjugation and submission to the ruling power, whether that be the eating body or the colonising nation, provides an articulation of the body politic at work in the imperial project. The philosopher Frantz Fanon, arguably the most influential anti-colonial thinker of the twentieth century, described the colonisation of countries like his native Martinique as a digestive process, in which European culture and values are foisted upon the palettes of a populace with no appetite for them. 'In the colonial context', he writes, 'the settler only ends his work of breaking in the native when the latter admits loudly and intelligibly the supremacy of the white man's values', but during the 'period of decolonization, the colonized masses mock at these very values, insult them, and vomit them up'.[252] Freedom from an imposed regime that claims dominion over the bodies and minds of its subjects, necessitates – for Fanon – a purging of the belly, puking up the poison you have been forced to swallow. Here mechanisms of ingestion, absorption and expulsion track meaning between individual, collective and political bodies in ways that register the corporeal impact of power and organised defiance.

The gut is not only an organ that stores and then violently evacuates, it is also a source of fun, humour and irreverence that can border on the revolutionary. In *The Wretched of the Earth* (1961), Fanon describes laughter as a powerful tool of resistance because to laugh 'in mockery' at 'Western values' is to refuse 'the violence' and 'the aggressiveness' through which those values are 'affirmed'.[253] Though fables like 'The Belly and Its Members' have enlisted the stomach in support of top-down paternalism, the cultural history of the belly also reveals more disruptive potentials. The maintenance of the dominant social order has, as explored in Chapter Three, often involved the careful regulation of the body's baser functions, especially those associated with consumption, digestion and defecation. To upend the established body politic might, as we have seen, involve looking to the bawdy, burping, slurping gut an alternative source of authority. Along similar lines, while narrow versions of national identity have been bolstered by the pleasures and anxieties associated with eating, there are other, less hierarchical, ways to imagine the work of the stomach. In *Dangerous Digestion: The Politics of American Dietary Advice* (2015) the environmental sociologist E. Melanie DuPuis argues that an alternative story might be found in our emerging understanding of the microbiome and the important role it plays in the health of the body. Once eaten, food, she notes, 'seems to be asocial', but in truth the 'fermenting that occurs in digestion is completely social, a collaboration between humans and the microbes they have domesticated, and which have domesticated them'. Instead of imagining the digestive process as a means by which the self conquers the non-self, we should think of it, she writes, as an example of 'how people can live as a part of the multitude of beings, with a body that is a collaboration with these beings'.[254] Rather than use the inner workings of the body to naturalise long-entrenched hierarchies, DuPuis holds up the gut microbiome as a utopian vision of what the world could be if we were only willing to follow its egalitarian example. Thinking more about the traction between bellies and politics, the next chapter returns to the history of the diet to consider what might be revolutionary about refusing to go on one.

11

The Revolting Gut

One of the most influential diet gurus in history was a funeral director. In 1863 William Banting published a short pamphlet, *Letter on Corpulence, Addressed to the Public*, that recounted his long struggle with excess weight and the regimen that eventually freed him from the terrible burden of 'Obesity'. The book was an immediate success, going into its fourth edition by 1869, with translations in French and German.[255] Its principles were so widely taken up that 'banting' remained a popular shorthand for dieting well into the twentieth century. Even today, Swedish slimmers describe themselves as trying to *att banta*, a common term for weight loss derived directly from mid-Victorian Britain. For readers, the appeal of *Letter on Corpulence* lay not only in the simplicity of the advice offered, but also in the relatable personal history that framed it. As he moved through middle age, Banting was perturbed to find that over time his waistline increased to the point that he was unable to 'attend to the little offices humanity requires without considerable pain and difficulty'.[256] Doctors told him to take up more exercise so Banting took to walking long distances through the city and rowing along the Thames every morning, activities that only served to further stimulate the appetite and failed to halt his steady weight gain. Eventually, having tried an intense course of Turkish baths and every 'physic' known to man, he found the solution to his problem from an unexpected source. During a consultation with William Harvey, an ear, nose and throat surgeon, regarding the progressive loss of his hearing, Banting was urged to cut carbohydrates from his diet. In place of bread, butter, milk, sugar, beer, soup and potatoes, he was to consume meat, fish and the occasional slice of dry toast. Over the next few months Banting shed forty-six pounds and felt his health restored to such a miraculous degree that he was inspired to share his story with the world.

Today advocates of diets like the Paleo (consume only what would have been available to our hunter-gatherer ancestors), the Atkins (eat as much protein as you like) and the Keto (reduce intake of starchy foods to such a degree that the body enters a state of 'ketosis' and begins burning fat for fuel) cite *Letter on Corpulence* as the foundational text for the low-carb movement and claim its author as an early pioneer in the battle against bread. Yet while it is possible to see this account as a triumphal tale of slimming success, a more ambiguous reading is also possible. From his characterisation of obesity as the most 'distressing' of 'all the parasites that afflict humanity' to the anxious insistence that his 'corpulence and subsequent obesity was not through neglect of necessary bodily activity, nor from excessive eating, drinking, or self-indulgence of any kind', it is clear that Banting was made to feel deeply ashamed of his body.[257] In one particularly moving passage he recalled the experience of negotiating the city as a fat person:

I am confident no man labouring under obesity can be quite insensible to the sneers and remarks of the cruel and injudicious in public assemblies, public vehicles, or the ordinary street traffic; nor to the annoyance of finding no adequate space in a public assembly if he should seek amusement or need refreshment, and therefore he naturally keeps away as much as possible from places where he is likely to be made the object of the taunts and remarks of others.[258]

Rather than read *Letter on Corpulence* as a record of marvellous dietary innovation, it might be better understood as a moving personal account of what it is like to inhabit a body that the society around you holds in contempt. What disturbs the hardworking Banting most, that his fleshiness might be interpreted as evidence of laziness, speaks to the way that fatness and morality seem to be fatally entangled with one another. This association has a long history. From Galen's assertion that 'it is truest of all that a full stomach does not beget a fine mind' to the early Christians who condemned corpulence as an impediment to spirituality, from the eighteenth-century moralists who worried at the impact of excess weight on the capacity for rational thought to the twentieth-century military recruiters who condemned modern man as too soft of both mind and of belly, body size has long been interpreted as a negative outward reflection of a person's character.[259] Weight gain was a dreaded 'calamity' for Banting not only because of the physical difficulties it might present, but also because 'one so afflicted is often subject to public remark'.[260] And conversely, losing weight meant reclaiming his place as a respected and valued member of society.

Which raises the question: what exactly are diets for? Borrowing the term 'gastropolitics' from the American food historian Frederick Kaufman, who uses it to describe how 'our understanding of virtue and vice, success and failure, has long been expressed in the language of appetite, consumption, and digestion', this chapter examines the complicated politics of dietary advice.[261] From the medieval physicians who prescribed foods based on their patients' humoral composition, to the seventeenth century where Roger Crab was advocating raw food as a path to God, to the Victorian vegetarians who touted the meat-free lunch as a cure for indigestion, the subject of diet has recurred throughout this curious history of the gut. Picking up some of these dropped threads, this chapter will examine the political resonances of the diet to better understand the muddying of corpulence with character that so tortured Banting. The first meaning of the word 'diet' given in the *Oxford English Dictionary* has nothing to do with food. Derived from the Middle English *dyete* and the Latin *dieta*, up until the early seventeenth century the term was often used to describe a way of thinking or a mode of life.[262] So in 'The Tale of Beryn', one of Geoffrey Chaucer's last *Canterbury Tales* published at the end of the fourteenth century, the merchant pilgrim optimistically imagines that 'Ech day our diete Shall be mery & solase', and John Lydgate's fifteenth-century poem *The Dietary* – which urged that 'Moderate fode gyffes to man hys helthe/ And all surfytys do fro hym remeve' – offered advice that was more moral than medical.[263] Though this particular usage eventually fell out of favour, today the word retains some of its original meaning: when we diet we are transforming our lives as well as our bodies, elevating control at the kitchen table as a supreme act of will capable of revolutionising every aspect of our existence, and imagining that by taming the belly we might become better people. At stake here is what Michel Foucault described as 'technologies of the self', namely the advocacy of individual responsibility and bodily discipline as essential to personal hygiene, civic virtue and the maintenance of the national body.[264] Along similar lines, this chapter examines dietary advice as not only a means of disciplining the body, but also a way to control or reform society. Determining what we should eat is also about articulating what kind of society we want to live in, what we want it to look like, on whom we bestow citizenship and who we wish to exclude.

Until now *Rumbles* has mainly thought about 'the gut' as an interior confederacy of liver, stomach, oesophagus, gallbladder, pancreas, small and large intestine, but the term is just as commonly used to describe an exterior feature of the body: the fleshy middle, the tummy, the paunch, the belly, the breadbasket, the bay window, the feed box, *kyte* (Scots), *bedaine* (French) and *bol'shoy zhivot* (Russian). Whatever you call it, today the abdomen is the target of a booming industry dedicated to firming and flattening it. Alongside books like *Flat Belly Diet!* (2008), *The Lose Your Belly Diet* (2016), *Zero Belly Diet* (2014) and *21-Day Tummy Diet Cookbook* (2014), it is now possible to buy supplements that are 'specially formulated' to reduce bloating, detoxifying teas that supposedly target belly fat and cold-pressed vegetable juices to take the place of stomach-expanding solid foods. Many of the most recent weight-loss plans call on cutting-edge research into the microbiome for scientific support. As we have seen, today scientists working in several different fields – endocrinology, metabolomics, physiology – are exploring the role that microbial cells in the intestinal tract might play in governing appetite, setting metabolic pace and regulating blood sugar levels.[265] Most attention has been paid, perhaps unsurprisingly, to the possible connection between the presence of certain bacteria and an individual propensity to obesity. Several studies have raised the tantalising possibility that the gut might hold the secret to easy and sustainable weight loss. In one experiment, undertaken by endocrinologists at Massachusetts General Hospital in Boston, a small group of patients classified as medically 'obese' were given capsules to swallow that contained the faeces of slim volunteers, to test whether transplanted gut bacteria from their skinnier compatriots had any impact on the metabolism of the test subjects.[266] These 'faecal microbiota transplants', a version of which we came across in Letchworth Garden City, were administered with the hope of increasing insulin sensitivity and creating a more diverse gut flora in the recipient. That the results of this study and others like it proved largely inconclusive did not prevent a flurry of wild speculation as to the miraculous possibilities of bacteriotherapy: the *Daily Mail* predicted that soon we will all be following 'personalised diet plans [. . .] tailored to the microbiome', the *Guardian* featured articles on 'Seven Ways to Boost Your Gut Health' and the rise of 'Super Poo', while the *Independent* reported that the secret to weight loss lay with 'testing your gut bacteria'.[267] Long maligned as the source of unwanted fat, the belly is now imagined as the key to achieving a slimmer physique.

At first glance it appears that the dissemination of research into the microbiome and the popularisation of ideas about microbial health may have helped to institute an approach to dieting far removed from the punitive regimes of the past. Out with tasteless sugar-free snacks, calorie counting and portion control, in with nurturing your biome, cultivating good bacteria and eating in harmony with the needs of the digestive system. However, on closer inspection this new health orthodoxy is underpinned by a familiar set of assumptions concerning the ideal body and the moral qualities required to achieve it. For one, a 'gut healthy' diet – like any other diet – involves closely monitoring and restricting consumption, exercising self-control and resisting indulgence. Most pressingly, the fat body must still be avoided at all costs. In 2016 newspapers reported with horror on a study which suggested that obesity might be contagious.[268] Biologists at the Wellcome Sanger Institute discovered that spores produced by bacteria in the gut could survive outside of the body and might be passed to those around us.[269] While the researchers responsible emphasised the benefits of their discovery – how it would make genome sequencing easier and deepen our understanding of gastric illnesses – what made headlines was the terrifying prospect that, as one article put it, 'spores of bacteria from the guts of fat people could spread to healthy individuals'.[270] Not only is the soft belly apparently an offence to the eye, it is also a storehouse of chubby bacteria that threaten to undermine the hard-won svelteness of others. The spectre of contagion raised here reveals how, even in seemingly neutral discussions of diet-induced bacteria like *Clostridium*, *Faecalibacterium* and *Eubacterium*, the fat body emerges as an object of disease and disgust.

This perception is, as cultural theorists Jana Evans Braziel and Kathleen LeBesco have argued, 'not natural', but rather a 'function of our historical and cultural positioning in a society that benefits from the marginalization of fat people'.[271] One of the sites where this marginalisation is felt most acutely is, according to many activists and theorists, in therapeutic settings.[272] Today to be overweight is to be subject to intense clinical scrutiny and even tacit disapproval, but fatness has not always been thought of as a medical problem or even as a problem at all. According to the doctrine of the four humours, body weight was dictated by an individual's peculiar balance of yellow bile, black bile, blood and phlegm. Those of a phlegmatic temperament, in whom the colder, wetter elements predominated, were thought to be generally plumper than those of, say, a more choleric nature who, being dominated by hot, dry forces, were more likely to be thin. Either way, at least until the late sixteenth century an individual's physique was most commonly viewed as an expression of the same interior forces that dictated everything from hair colour and face shape to their emotional style and dominant personality traits. Fat was, in other words, not in itself necessarily always a problem. In some it might indicate a humoral imbalance, but in others it might be taken as evidence of good health; physicians were as likely to warn against eating too little as they were to inveigh against eating too much.

This began to change, as food historian Ken Albala has argued, around the middle of the seventeenth century when the medical profession first began to establish 'fat' as a distinct pathology. The reasons for this shift were complex: new chemical and mechanical discoveries in the field may have had a part to play, or perhaps, as Albala conjectures, it was simply the 'first time that enough people in one area had enough expendable income to seek a cure, making obesity a lucrative medical specialty and the object of medical controversy'.[273] Whether the result of deepening knowledge of the body or evidence of blatant profiteering, by the end of the eighteenth century corpulence had been successfully rebranded as not only a problem, but more importantly as a problem that could only be solved by the intervention of a knowledgeable physician. Where the belly had once been open to multiple conflicting interpretations – evidence of sinful gluttony or a fantastic display of wealth, an object of erotic fascination or a freakish spectacle – increasingly it came to be seen solely through the eyes of medical men.

Such was the medical community's growing obsession with portliness that by the publication of *Cursory Remarks on Corpulence; or, Obesity Considered as a Disease* in 1816, its author William Wadd was able to fill over a hundred pages with the proliferation of theories as to its 'causes and cure' as advocated by his fellow doctors. The advice gathered by Wadd, who was a member of the Royal College of Surgeons and one of George IV's personal physicians, begins with the commonplace injunction to eat less and exercise more, but quickly moves out into less familiar territory.[274] One doctor prescribes drinking Castile soap mixed with lime water, another recommends consuming only over-salted meat and water to 'render fat more fluid', while another holds resolutely to the ancient practice of bloodletting.[275] Horse-riding, hot baths, sleeping less and talking faster are also posited as potential aids to weight loss. Not only does there appear to be little consensus as to the best method for removing fat, but the experts canvassed in *Cursory Remarks on Corpulence* also seem at odds as to what kind of substance fat is and why it seems to build up in certain areas of the body. We learn from a 'celebrated' anatomist that, at the cellular level, fat is the 'connecting medium' holding the body together;[276] others point to the omentum – the fleshy abdominal area in front of the intestines – as the 'common root' of all fat; the Italian biologist Marcello Malpighi[277] thought its spread was aided by some kind of 'glandular apparatus'; the famous Dutch physicians Herman Boerhaave[278] and Gerard van Swieten[279] 'were of the opinion that fat is deposited from the blood by its slower circulation'; while according to popular opinion it was obvious that the stomach was the sole 'manufacturer of fat'.[280] Though no coherent clinical picture emerges in his book and very little is said of exactly how 'excess' fat might impact the functioning of different parts of the body, Wadd is clear on the moral implications of corpulency. Drawing a line between physiology and character, he observes that because of 'the general pressure on the large blood vessels, the circulation through them is obstructed' and hence 'we find the pulse of fat people weaker than in others, and from these circumstances also, we may easily understand how the corpulent grow dull, sleepy, and indolent'.[281] The rate at which blood is being pumped around the body can, it would appear, sometimes reveal a great deal about a person's character.

The problem with attempting to cure obesity, according to Wadd, is that the afflicted cannot be counted upon to follow the sound advice issued by their knowledgeable physicians. Though the simplest solution to excess weight is to adopt a plain and moderate diet, accompanied by regular exercise, it often proves impossible for the patient who has cultivated habits 'connected with great inactivity of body and indecision of mind' to muster the 'continued perseverance' required.[282] Corpulence was, then, perceived as not only a symptom of immoderation or gluttony, but it could also be interpreted as a sign of mental inferiority. Early modern physicians attributed this to the way that clots of fat impeded the flow of animal spirits through the body, dulling the senses and sapping vitality.[283] In the eighteenth century, doctors were more likely to describe the obese body as prone to laxity – loose skin, inelastic tissue and distended vessels – but the conclusion drawn was the same. Namely, that it constituted an outward expression of inner traits, primarily laziness, slow wits and lack of willpower.

By the time Wadd met an untimely and dramatic end in 1829 – killed jumping from a runaway carriage while on a family holiday touring the south of Ireland – the concept of the 'will' as a key measure of character was well established. Though this idea has deep roots in Western culture – think of the importance of free will in Christian theology – it was not until the nineteenth century that it emerged as the dominant way of thinking about everything from morality and civic duty to race, gender and sexuality. This powerful cultural narrative was popularised by bestselling books like Samuel Smiles' *Self-Help* (1859), which celebrated the virtues of hard work, thrift and perseverance. The secret to success in life lay, according to Smiles, with the vigorous cultivation of 'self-culture'; the lesson being that moral character is shaped not by 'riches and rank', but by qualities like self-sufficiency and self-discipline. Autonomy was essential, he insisted, because where 'help from within invariably invigorates', the help proffered by others is 'often enfeebling in its effects' and in fact 'everything depends upon how he governs himself from within'.[284] This call for greater self-governance was very much in tune with the values of industrial capitalism, which was similarly invested in the promotion of a kind of atomistic individualism grounded in ideals of responsibility and self-resilience. With the result that where a rounded figure might have once signalled wealth and success, in nineteenth-century Britain it signified only one's failure to live up to the exemplar of the productive, disciplined citizen.

Diet literature from the period, books like Banting's *Letter on Corpulence* as well as countless magazine articles and instructional pamphlets, emphasised the need to establish proper governance so that, as the historian Sander Gilman put it, the 'will becomes that which is healed by the dieting process and enables the rational mind to control the body'.[285] This advice was grounded in what might be described as a juridical model of the self. Following the Cartesian separation of mind and body, this charged the logical, reasoning higher facilities with bringing order to the irrational demands and pleasure-seeking desires of the hungry belly. Towards the end of the nineteenth century, the image of what constituted a well-managed body and what did not became the subject of an emerging scientific discipline. In 1884 attendees of the International Health Exhibition (IHE) in South Kensington – which we visited briefly in Chapter Seven – were encouraged to try out the world's first Anthropometric Laboratory. Stationed alongside promotional displays of Pears Soap, model sanitation systems and a hydrotherapeutic clinic, this temporary laboratory was the work of Francis Galton, a psychologist, psychometrician and cousin of Charles Darwin, who pioneered the application of statistical methods to the study of human development and difference. Visitors to the pop-up test centre were introduced to new methods of measuring and tabulating their physical characteristics, then encouraged to submit information such as arm span, visual acuity and head size to the growing bank of data. Hugely popular with the thousands that frequented the exhibition, it employed techniques that had been developed over the course of a large-scale comparative study of the nation's physical features, undertaken by the British Association for the Advancement of Science (BAAS) between 1875 and 1883.[286] Information gathered by the BAAS found practical application in eugenics, a science developed by Galton that identified and propagated traits deemed useful to the further evolution of the species. Man's abilities are, he argued, derived from his inheritance, and as such each generation has 'enormous power over the natural gifts of those that follow'.[287] Endorsed by the medical community and popularised through events like the IHE, the survey also helped to create the first statistically 'normal' body, defined by measurements like height, weight and waist size, that was swiftly instituted as the acceptable median to which all must aspire.

From the beginning, the practice of eugenics relied heavily on lay participation, cultivated through population surveys, data collection and self-reporting. In an advertisement for the Anthropometric Laboratory potential visitors are encouraged to attend on the basis that, having had their body carefully observed and measured, they may receive a 'timely warning of remediable faults in development' (Figure 31). While evolution through natural selection was slow and unpredictable, eugenicists pointed to the role that an individual could play – in their choice of mate, in the education of their children, in the maintenance of a healthy weight – in the future progress of the race. The fat belly, now subject to exacting and authoritative methods of quantification, signified an unacceptable deviation from the physiological ideal.

The idea that it might be possible to establish what a 'normal' body looks like was first seriously pursued in the early part of the nineteenth century by figures like the Belgian scientist Adolphe Quetelet. An early pioneer of what he described as 'social physics', he was one of the first to apply insights from probability and statistics – usually confined to fields like mathematics and astronomy – to the study of the social world.[288] In his treatise on the subject, *Sur l'homme et le développement de ses facultés, ou Essai de physique sociale* (1835), published in English as *A Treatise on Man and the Development of His Faculties* (1842), Quetelet extolled the virtues of the 'average man'. By collecting and comparing data from individuals it was possible, he argued,

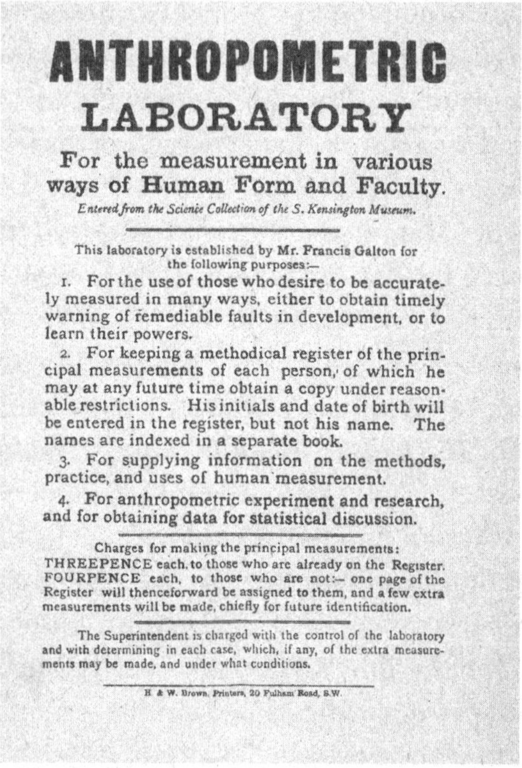

ANTHROPOMETRIC
LABORATORY

For the measurement in various ways of Human Form and Faculty.

Entered from the Science Collection of the S. Kensington Museum.

This laboratory is established by Mr. Francis Galton for the following purposes:—

1. For the use of those who desire to be accurately measured in many ways, either to obtain timely warning of remediable faults in development, or to learn their powers.

2. For keeping a methodical register of the principal measurements of each person, of which he may at any future time obtain a copy under reasonable restrictions. His initials and date of birth will be entered in the register, but not his name. The names are indexed in a separate book.

3. For supplying information on the methods, practice, and uses of human measurement.

4. For anthropometric experiment and research, and for obtaining data for statistical discussion.

Charges for making the principal measurements:
THREEPENCE each, to those who are already on the Register.
FOURPENCE each, to those who are not:— one page of the Register will thenceforward be assigned to them, and a few extra measurements will be made, chiefly for future identification.

The Superintendent is charged with the control of the laboratory and with determining in each case, which, if any, of the extra measurements may be made, and under what conditions.

H & W. Brown, Printers, 20 Fulham Road, S.W.

Figure 31: Pamphlet advertising Galton's Anthropometric Laboratory at the International Health Exhibition

to delineate 'average bodily proportions' and in doing so set the 'limits within which variations [. . .] must be kept in check, in order not to shock the taste'.[289] In other words, setting the 'average' made it easier to distinguish between acceptable and unacceptable bodies. The metric that Quetelet devised involved dividing weight in kilograms by the square of height measured in metres to produce a ratio; a ratio now better known as the Body Mass Index (BMI). As the rubric through which 'obesity' is defined, the

BMI wields immense medical and cultural power, but its origins reveal its limitations as a useful indicator of health. Most problematic, Quetelet formulated his ideas based on measurements taken from French and Scottish samples, meaning that the 'average man' was created using only information provided by white Western Europeans.[290] In addition to producing dangerous clinical blind spots – the overdiagnosis of some groups and the underdiagnosis of others – a statistical norm based on whiteness and maleness also serves to reinforce long-established structures of power. A fat belly can disqualify its owner from wielding power in society, on the grounds that they are morally deficient, mentally inferior, or simply lacking in the necessary self-control. And so, while Quetelet and eugenicists like Galton presented their methods as firmly grounded in the unassailable principles of scientific objectivity, it is telling that their statistical models tended to exactly reproduce rather than challenge the status quo.

When Galton set up his makeshift laboratory at the IHE, willpower was viewed as primarily the preserve of white, educated men and it remained the basis on which everyone else – women, non-white, working class – was deemed to be unable to govern themselves. Such bodies are not so easily absorbed in the imaginary of the 'national body' and thus present difficulties to state authorities trying to create a coherent image of the ideal citizen. Considered from the vantage point of political institutions, diet arguably enacts a kind of coercion, a means of disciplining not only the gut but also the populace. Renaissance moralists, for instance, took the question of diet very seriously because they thought that a well-fed

populace was easier to govern and more disposed to obedi-
ence. Ensuring not only that everyone had enough to eat,
but also that they were discouraged from overindulging in
alcohol, was – as countless fourteenth- and fifteenth-century authors attest to – a remarkably effective means of
social control. As Ken Albala argues in *Eating Right in the
Renaissance* (2002), this was usually understood in terms
of humoral balance, meaning that a 'gluttonous people fed
on choleric meats is ready to take offence at the slightest
cause; melancholics invent false dangers to fret over; the
phlegmatic loses all vigor and motivation. And of course
a pusillanimous population is a threat to national security.'[291] If the history of diets is, at least in part, the history
of governance and the rule of the few over the many, then
is the fat body a profoundly disorderly one? An etching
by the eighteenth-century Scottish satirist Isaac Cruikshank, *The Doctor and the Unruly Patient* (1797), seems to
suggest so. A frustrated physician chastises his grinning
patient for growing 'as fat as a porpoise', who answers that
he certainly has not 'followed his prescription', as he has
chucked it out of the window (Figure 32). This gleeful
insouciance suggests more than a thick-headed disinclina-
tion to do what is necessary to stay in good health; rather
the patient's refusal to submit to the authority of his physician gives some insight into the radical possibilities of
being 'unruly'.

Recently, fat activists and theorists have argued that
rather than viewing fatness in terms of aesthetics or as
a medical condition, we should recognise it as a political
issue. If, as Kathleen LeBesco has proposed, we think of
'revolting' bodies in terms of their ability to overthrow

Figure 32: *The Doctor and the Unruly Patient*
by Isaac Cruikshank, 1797

authority, to rebel and protest, then fatness could be conceived of as a way of subverting dominant discourse and upending established authorities.[292] Less easy to assimilate into the ideal of the national body – strong, self-sufficient, disciplined – the soft belly exposes the inequities and injustices that underpin modern society. Fat bodies are often described in the language of excess and inefficiency, viewed as less productive and therefore less valuable to capitalism. Nowhere is this more apparent than in the well-documented weight bias that women encounter in the workplace. While body size does not appear to have much impact on the career progression of men, studies have shown that women perceived to be overweight are consistently penalised in terms of hiring, salary and promotion. The problem also appears to be getting worse. A study from Harvard University analysed the attitudes of millions of participants between 2007 and 2016 and found that while biases based on sexual orientation and race were on the decline, in contrast anti-fat bias had risen by 40 per cent.[293] Not much seems to have changed since William Banting wrote his *Letter on Corpulence* in 1863: fleshiness is still interpreted as proof of indolence and the vitriol directed at bodies that exceed what is 'normal' reveals the toxicity of such a measure. What is peculiar is the gendered dynamic at play in this kind of workplace discrimination, which is significant as it suggests that fatness might have different cultural resonances, political meanings and economic implications for women and men. Moving on from the question of body weight but sticking with the gut as a historically unruly organ, the final chapter of *Rumbles* enquires into the digestive politics of gender.

12

The Gender of the Gut

In 1899 an Australian newspaper carried an advertisement for a laxative that, on the face of it, seemed to have very little to do with the promotion of regular bowel movements. Instead of touting the product's superior value and unique purgative properties, the advert derided the 'mannish mode and airs' of the 'new woman'.[294] Who was this woman and what made her so novel? As the Victorian period ended the so-called 'New Woman' burst forth into the world. Unlike her dowdy domestic forebears, she was educated, employed, politically active, unmarried, fond of bicycles and prone to wearing trousers. Known in Japan as an 'atarashii onna' and in Germany as a 'Neue Frau', these emancipated women provoked a widespread cultural panic over changing gender roles, the erosion of respectable feminine behaviour and the decline of the traditional family. The charge most frequently levelled at these new women, who were seeking a broader sphere of action in the world, was that to agitate for such a change was to struggle fruitlessly with the sensible restraints set by human biology. This was certainly the approach taken by the odd Australian advertisement for a product called 'Bile Beans', which proclaimed that 'nature is against' those seeking to do more than keep a home and raise a family, because the female 'internal organism can stand less strain and needs a great deal more attention than does the male'.[295] Here, as is almost always the case, sexual difference implies male superiority and female weakness, a handy biological justification for the social, political and economic domination of the former over the latter. What is distinctive, however, is the insistence on the gut as the chief cause of feminine frailty.

This particular advertisement implies that women are more prone to digestive distress than men and yet at various points on this tour through the curious history of the belly the opposite has also been true. Think of the young noblemen of the Renaissance learning to control their bodily appetites or the nervous poets of the Romantic era beset by 'blue devils' or the overworked clerks of late-Victorian London tortured by bouts of painful dyspepsia, the troubled belly has just as often been associated with men as with women. Even a brief glimpse at the long view of the cultural history of the gut reveals, then, how constructed any seemingly natural understanding of its relationship to gender is. Published at the turn of the twentieth century, a period of great social change that saw women agitate for the vote and enter the workforce in ever greater numbers, the Australian Bile Beans advert reflects the anxieties generated by such a shift and reveals once more where the intestinal world intersects with political life. Such anxieties were not confined to women; this period also saw established models of masculinity come under strain as a new culture of the body emerged that emphasised regularity as a test of true manhood. Taking a deep dive into the dietary cultures of the early twentieth century, this chapter examines the politics of peristalsis and asks what digestion can tell us about the history of gender. Along the way it picks up on two of the constipation-obsessed reformers introduced in Chapter Nine, spends time with the first celebrity bodybuilder and goes on hunger strike with militant suffragettes, all in an effort to better understand the gender of the gut.

The makers of Bile Beans, having characterised the stomach as particularly vulnerable to disorder and upset, marketed themselves almost exclusively to women. Beginning in the late nineteenth century and running well into the 1960s, advertisements for the laxative promised to not only restore regularity, but in doing so solve menstrual problems, improve complexion, brighten eyes and encourage weight loss. A poster from the 1940s is typical: a glossy-haired woman in high heels stares anxiously at a set of bathroom scales, but happily she need not worry as her nightly dose of Bile Beans has kept her 'healthy, happy and slim' (Figure 33). Not only did physiology predispose women to constipation, but sluggish bowels also threatened to undermine the successful performance of femininity. Inattention to digestion might lead to a thick waist and dull complexion, jeopardising marriage prospects and forcing women to adopt unattractive 'mannish' ways to compensate. While it is no surprise to find women in history being judged on their appearance and warned that their future happiness depended upon the cultivation of good looks, it is significant to find the movements of the digestive tract implicated in this familiar cultural narrative.

Bile Beans promised to maintain the dominant social

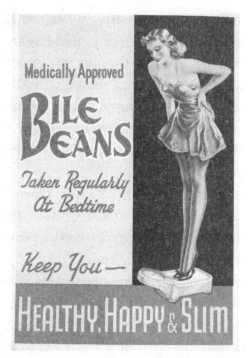

Figure 33: Advertising card for Bile Beans

order by managing the digestive process. A bold ambition that was bolstered by its origin myth, which held
that a chemist by the name of Charles Forde had synthe-
sised the unique preparation from a mysterious vegetable
substance hitherto known only to the Aboriginal Austral-
ians. Though a 1903 investigation by the *British Medical
Journal* and a court case a few years later concluded that
they were, in truth, composed mainly of rhubarb, liquo-
rice, cascara and menthol, the gelatinous beans remained
popular well into the twentieth century and helped to
establish constipation as a distinctly female ailment.[296]
Recent studies suggest that women are more likely than

men to suffer from costiveness – or at least more likely to seek help for it – and the NHS identifies menstruation, childbirth and the menopause as common contributing factors, but the digestive culture that Bile Beans helped to create interpreted this disparity as evidence of a building crisis in modern womanhood. William Arbuthnot Lane, the surgeon who we encountered worrying over his patients' colons in a previous chapter, was convinced that women were at far greater risk of intestinal stasis due to a combination of physiological and social factors. Women were, he insisted, not only physically weaker, anatomically compromised and constricted by unnecessary corsets, they were also hampered by prudishness and social convention. This reluctance to discuss toilet matters could lead to all manner of trouble. Waste backed up in the digestive system sapped vitality and inhibited the appetite. The constipated woman was, according to Lane, a rather unlovely creature. Slowly poisoned by the toxins amassing in the lower intestine, her appearance deteriorates – the 'buttocks become flat and flaccid' and breasts 'waste and flop downward' – and the patient becomes increasingly irritable and unhappy.[297] Cursed with fading looks and a bad attitude, such women were unlikely to marry and would remain childless; and with constipation on the rise in the developed West, this could, he warned, lead to falling birth rates, jeopardising productivity and threatening the integrity of the national body.

While Lane advocated a rather drastic surgical solution to the problem of chronic constipation – the removal of a large section of the colon – less extreme treatments were available. One could, for example, simply begin the

Here is what that dread disease
CONSTIPATION
will do if neglected

1. Tire you out.

2. Sap your reserve energy.

3. Put wrinkles in your face and gray hair in your head.

4. Cause pimply skin.

5. Cause bad breath.

6. Give you headaches.

7. Dull the brain.

8. Cause stomach disorders.

9. Weaken your entire system.

10. AND LEADS TO OVER 40 OTHER SERIOUS DISEASES!

Don't Dare to Neglect
CONSTIPATION!

These 2 women are the same age

One has the bloom of youth.

The other is wrinkled, gray, careworn, far older than her years.

Figure 34: Leaflet advertising Kellogg's All-Bran, 1930

day with a bowl of wholesome breakfast cereal. Throughout the early decades of the twentieth century, branded cereals were promoted as not only cures for constipation, but also the secret to beautiful, vital womanhood. A 1930 advertisement for All-Bran, for instance, features two women, one who has retained the 'bloom of youth' and another who is 'wrinkled, gray, careworn, far older than her years'; while one has attended to her bowels, the other endures the 'dull brain', 'pimply skin' and 'bad breath' that comes from neglecting digestive health (Figure 34). More than canny marketing, these claims were grounded in the popular medical culture of the period, specifically in the transatlantic exchange of ideas on the role of the gut in overall wellbeing. Exemplified by John Harvey Kellogg's 1907 trip to London to attend Lane's classes on gastrointestinal surgery at St Bartholomew's Hospital, this sharing of ideas helped to establish the consensus over the threat of autointoxication explored in Chapter Nine. The American approach to digestive health distinguished itself from the British in one key respect: while Lane emphasised its relationship to temperament, reformers across the pond were exploring its spiritual dimensions.

John Harvey Kellogg managed, with his younger brother Will, a popular health resort in Michigan called the Battle Creek Sanitarium: an impressive complex that featured a hospital and nursing school, as well as guest accommodation and facilities that included a golf course, a swimming pool and a dining room. The resort was operated according to the health principles of the Seventh-day Adventist Church, a Protestant denomination to which the Kellogg brothers belonged. Born out of what came to be known as the Second Great Awakening, a religious revival that swept through the United States in the first half of the nineteenth century, Seventh-day Adventism preached the imminent and literal return of Christ to earth, after which the righteous would finally ascend to heaven. What distinguished it most starkly from other Christian denominations was the emphasis it placed on health matters. As Ellen G. White, proclaimed prophet and co-founder of the movement, explained: because 'spiritual vigour' depends on 'physical strength and activity', service to God must begin with the health of the body.[298] Inspired by divine visions and her interpretation of biblical teachings, White developed a model of health that emphasised hygiene, moderate diet and chastity. Believers were instructed to exercise regularly and preferably in the open air, abstain from alcohol, tobacco and caffeine, and follow a vegetarian diet rich in nuts and pulses. This wholesome regimen was designed to ward off the temptation to indulge in less wholesome pursuits. Most dangerous among these was the sin of 'self-pollution' or masturbation, an activity that Adventists held responsible for a wide range of physical, mental, emotional and spiritual problems. To Kellogg, masturbation constituted a form of self-abuse that exhausted the body's supply of nervous energy, inviting disease and – in especially severe cases – courting insanity. And any life-long commitment to resisting these dangerous urges must begin with breakfast.

The kitchen of the Battle Creek Sanitarium was a site of culinary experimentation and innovation. Throughout the final years of the nineteenth century, the Kellogg brothers and John's wife Ella developed and marketed a range of patented health foods including the first peanut butter, meat substitutes and, most famously, cornflakes. These were made using a baking process described as 'dextrinisation' that used a very hot oven to break down the complex starches of wheat and corn into the simple sugar dextrose, resulting in a grain that was easier to digest and therefore less likely to cause gastric upset. Starting the day with a bowl of plain-tasting, high-fibre flakes was – for Dr Kellogg – the key to both physical wellbeing and moral fortitude. As was touched upon in a previous chapter, Kellogg was an early adopter of germ theory and was among the first to popularise the idea of intestinal flora as playing a pivotal role in maintaining the health of the body. In his 1918 study *Autointoxication or Intestinal Toxemia*, Kellogg described the methods employed at Battle Creek to 'reform' the bacteria of the gut, which ranged from dietary change and fasting to enemas carried out with gallons of water and pints of yogurt. A clean colon meant a clean mind because plain foods and regular bowel movements were the best insurance against the siren call of self-pleasure. It was, Kellogg advised in his *Ladies' Guide in Health and Disease* (1883), especially important that young women learn to discipline their animal appetites, because masturbation was a leading cause of not only moral degeneracy, physical deficiency and insanity, but also infertility. A nationwide effort was needed to improve the fitness of childbearing women and in 1914 he established the Race Betterment Foundation, an organisation that promised to address the 'present state of physical degeneracy [. . .] by means of attention to the laws of heredity'. Committed to the principles and practice of eugenics, Kellogg believed that the Western world was in a state of racial decay that could only be arrested by encouraging the fit to reproduce and preventing those with negative physical or intellectual traits from doing so. It was therefore essential that the right type of girls – namely white Anglo-Americans – be trained in the rudiments of health. Drawing a direct line from good digestion to sexual reproduction, Kellogg hailed cornflakes as part of the solution to the pressing problem of national degeneration.

He was not alone in encouraging women to maintain healthy bowels and throughout the early decades of the twentieth century female fitness was consistently framed in terms of not only personal responsibility, but also as a duty to the race. Natural beauty and a slim figure were essential, but women were also instructed to pay particular attention to diet and digestion when attempting to conceive. This message was imparted in the marketing of products like Bile Beans, in popular magazines like *Woman's Own* which was first published in 1932 and through dedicated organisations like the Women's League of Health and Beauty. Established in 1930 by Mary Bagot Stack, a physical educationalist who had developed a system of rhythmic gymnastics after travelling in India and observing the practice of yoga, it was originally run out of a YMCA on Regent Street in London. The league's mix of dance, callisthenics, exercise classes and slimming support proved hugely popular, and by 1937 it boasted 166,000 members with centres established around Britain. Its aim, in common with Battle Creek, was to promote 'racial health' by physically preparing women for motherhood. Exercises were geared towards the abdomen and pelvic region, areas to be perfected for the sake of beauty and strengthened in preparation for childbirth.

Though Bagot Stack warned against the lure of faddish diets, members of the league were encouraged to maintain a slim physique and those deemed to be 'abnormally fat' were instructed to starve themselves until they reached a more acceptable size. This chimed with the popular diet culture of the period, in which manuals like Frederick Annesley Webster's *Slimming Day by Day* (1933) and Milo Hastings' *Reducing and How* (1930) advised female readers on how to rid themselves of 'unsightly' flesh and cultivate the flawless feminine form. During the interwar period, as historian Ana Carden-Coyne has noted, with the growth of the fashion industry and the availability of cosmetic surgery, women were increasingly expected to strive not for normality, but perfection.[299] Laxatives like Bile Beans offered female consumers one way to meet this expectation. As we have seen, the 1920s and 30s was the gilded age of purgatives, when hundreds of branded evacuants flooded the market and an increasingly broad range of health issues were attributed to the malign influence of backed-up bowels. For women, achieving and sustaining the ideal slender figure meant ridding oneself of weighty internal matter.

The popular digestive culture that emerged throughout the early part of the twentieth century also addressed itself to men, but it did so on slightly different terms. While constipation was a problem for women because it threatened their beauty and childbearing capacity, for men sluggish bowels, along with the thick waist that sometimes accompanied them, were read as signs of weakness and effeminacy. In a period characterised by improved living standards, cheap food and sedentary jobs, concerns began to be raised as to the effect of modern convenience on public health. According to historian Ina Zweiniger-Bargielowska, men came under pressure to lose excess weight because a 'normative body' was increasingly viewed as 'an asset in the public world of work' and 'a symbol of national efficiency'.[300] As Europe rumbled towards the outbreak of the First World War and the so-called 'Scramble for Africa' gathered pace, the question of physical fitness became increasingly tangled up with Britain's imperial ambitions. According to Edwardian health celebrities like Eustace

Hamilton Miles, regular exercise was 'the duty of every member of the Empire'. Framing fitness and health as the allegiance owed by each citizen to their country, Miles looked forward to the day when Britain might rule over what he described as an 'Empire of Well-Being', where 'wherever we go and rule, we shall set the example of real Health'.[301] Grounded in a vision of masculinity defined by self-discipline, strength and sportsman-like fair play, in this colonial fantasy healthy bodies confirm Britain's superiority and justify the country's rule over other nations.

First known to the public as a champion real tennis player, who took home the silver medal in the 1908 Olympics, Miles really made his name as an outspoken champion of diet reform and as the proprietor of the best-known vegetarian restaurant in London. Located at the western end of Covent Garden, the Eustace Miles Restaurant had, according to one review, the 'suggestion of the gymnasium' in its high ceilings and polished wooden floors, and its owner was keen to capitalise on this association.[302] Adverts for his establishment placed emphasis on the suitability of its 'well-balanced non-flesh meals' for the 'Athlete' in training, exercise classes were held in one of its upstairs rooms, and on the way out the door customers could purchase a packet of 'Emprote' – a special protein preparation for 'cyclists, athletes and swimmers' – or choose a title from the stock of improving literature piled by the door. Alongside his own sporting career, Miles also ran exercise classes and published books on personal fitness. A number of these were directed at boys and young men: *The Eustace Miles System of Physical Culture* (1907) was a drill book for children containing a carefully gradated series of physical exercises and *Fitness for Play and Work* (1912) was prefaced with an introduction from Robert Baden-Powell, the founder of the Scouting movement. These books were intended as manuals for disciplining the male body through both a regimented programme of exercise and by the adoption of a carefully calibrated and entirely meat-free diet. Using his own sporting victories as proof of the efficacy of his dietary advice, he claimed to have won tennis matches fortified by only salad, Hovis bread and a very weak cup of tea. Wholesome, natural and moderate, the flesh-free diet was the best way to fuel the active body because vegetables and grains could be processed quickly into energy, while meat tended to overload and clog the digestive system. Where early vegetarian reformers had emphasised the cruelty or moral corruption inherent in meat eating, later advocates like Miles focused on the dietary deficiencies of animal products and on what he described as the 'bodybuilding power' of vegetables and legumes.

Miles was representative of a particular masculine ideal characterised by a slender physique and temperate habits, but there were other models available. Chubby men were also encouraged to aspire to the hard, toned physique of bodybuilders like the showman Eugen Sandow (Figure 35). Born in Prussia in 1867, as a young man Sandow made his living as part of a touring circus before making his name as an internationally renowned strongman and a paragon of male beauty. He honed his musculature according to the dimensions of classical Greek sculptures, adopting the recognisable poses of famous statues in photographs, before popularising his methods through dedicated

Figure 35: Photograph of Eugen Sandow collected
in his book *Body-Building; or, Man in the Making:*
How to Become Healthy and Strong (1904)

gymnasiums and in *Sandow's Magazine of Physical Culture*
(1898–1907). Like personal trainers today, Sandow emphasised the need to cultivate healthy eating habits along with
brawny arms and a six-pack. He insisted that not only
did a man's vital power come from good digestion, but
also that the intestines simply worked better when the
stomach was toned and free of fat. Another fitness guru,
the American Bernarr Macfadden, was so convinced of
the link between gastric dysfunction and effeminacy that

he published *Constipation: Its Cause, Effect and Treatment* (1924) to encourage men to cultivate better internal resilience. Macfadden, who changed his name from 'Bernard' to 'Bernarr' because he thought it sounded like the manly roar of a lion and authored books with titles like *The Virile Powers of Superb Manhood* (1900), argued that indigestion represented a failure of masculine virility. The soft and gassy urban man was, he warned, unlikely to make a happy marriage, produce robust children or contribute to the world in any kind of productive fashion.

It may seem strange to find strapping figures like Sandow and Macfadden, who were famed for their superhuman strength, fretting over the frequency of men's bowel movements, but the gut has long been imagined as a site of peculiarly masculine energy. To 'have guts' is to be tough, spirited and courageous; to go with your gut, to follow a gut feeling, to act in a way that takes guts, is to demonstrate a certain strength of character. And it is telling that all the examples cited by the *Oxford English Dictionary* to illustrate the usage of this colloquialism refer only to men. Around 1914 a Professor of Clinical Medicine at Ohio State University coined another term for this manly inner resolve: 'intestinal fortitude'. Travelling back to his office after coaching the college football team, John W. Wilce described how his thoughts of sport had mingled with the lecture he was about to deliver on physiology and brought the new expression to mind. It caught on quickly and soon took on special resonance as young men began to be called up to fight in the First World War. The outbreak of war in Europe and the increasing likelihood of overseas combat led the US military to question whether the nation was fighting fit. Attention was focused, as Ana Carden-Coyne has uncovered, on the symbolic and literal gut. 'Waging war', she writes, 'was a test of masculine ability, and a test of inner resolve as it manifested in the outward shape of the stomach's flesh', so that a fat belly came to represent both a physical and moral failure.[303]

Away from the battlefield, the stomach has also been consistently gendered as masculine in the business world. During his tenure as President of the United States, Donald Trump regularly claimed to be in possession of a remarkable and never-failing gut instinct. In an interview with the *Washington Post* from 2018, in which he blamed the Federal Reserve's decision to raise interest rates for causing mass layoffs, Trump insisted that 'my gut tells me more sometimes than anybody else's brain can ever tell me'.[304] It may go some way to explaining the Commander in Chief's notorious unwillingness to read briefing documents, but his faith in the interpretive power of the gastrointestinal system began as a business strategy. In *The Art of the Deal* (1987), a ghost-written bestseller that was part memoir and part business self-help, Trump attributed his success as a property developer to having always trusted his gut. Underpinning this boast is an image of the stomach as possessing special knowledge about the world, information that is only accessible to those brave enough to ignore the cautions of the intellect and forge a new path. This trope extends beyond the omnipotent fantasies of US presidents. In modern Japan, for instance, businessmen and politicians are also encouraged to follow their gut instincts. As political philosopher Robin M. LeBlanc has discovered, in Japan this culture of the belly helps to justify women's continued exclusion from positions of power, as they are considered less astute readers of the viscera and therefore constitutionally unsuited to the demands of public life.[305]

Though the feminine gut has typically been coded as fragile in contrast to the inner strength and killer intuition embodied by the masculine belly, it is arguably women who have most successfully harnessed digestion as a political strategy. During the long campaign for the vote in Britain, imprisoned suffragettes were subjected to force-feeding at the hands of prison doctors, a practice that involved restraining the prisoner while forcing a rubber tube into her mouth or nose, then pouring mixtures of milk, eggs or other liquid foods into the stomach. This was an incredibly violent experience, often resulting in broken teeth, bleeding, vomiting and choking, and one that was justified as the only sensible course of action when a prisoner refused to eat. The first suffragette to go on hunger strike was Marion Wallace Dunlop, a Scottish artist and author who was active in the Women's Social and Political Union (WSPU). Established by Emmeline Pankhurst, the WSPU was a militant organisation dedicated to winning women the vote through direct action that included smashing shop windows, burning stately homes, and eventually bombing several public buildings including Westminster Abbey. Operating under the slogan 'deeds not words' – a sly rebuke to the softly, softly approach of the 'law-abiding suffragists' organised by Millicent Garrett Fawcett – they cut telephone lines, spat at police, and engaged in other acts of violent resistance. In 1909 Wallace Dunlop was convicted, along with a male accomplice, of defacing the House of Commons and was sentenced to a month in prison. Details of the case were circulated in the national press: *The Globe* reported that the 'female defendant went to St Stephen's Hall, on June 24, and with a large rubber stamp impressed a notice on the wall of the building', which was recorded by the *Morning Post* as having read 'Women's deputation [. . .] Bill of Rights. It is the right of all subjects to petition the King. All commitments and prosecutions of such petitioning are illegal.'[306] After refusing to pay a fine of £5, Wallace Dunlop was sent to Holloway Prison in north London, where the authorities refused to classify her as a political prisoner. In protest she stopped eating and undertook a fast that lasted ninety-one hours, finishing only when she was released early as the prison management feared she might die otherwise. Following this success, the hunger strike became a key strategy for the WSPU and a big problem for the government. Early release meant capitulating to the demands of a militant group but allowing female prisoners to slowly starve themselves to death was politically toxic. Force-feeding, a practice that had long been employed by asylum doctors, was the answer.

Between 1909 and 1914, suffragettes imprisoned in facilities across Britain were held down, placed in restraints, their mouths prised open, and tubing forced down their throats. Sometimes the liquid nutriment was introduced through other orifices – the nose, rectum and vagina – defiling acts that many likened to being raped. To draw attention to these brutalities, the WSPU circulated leaked letters written from inside prison that contained graphic descriptions of force-feeding and condemnations of the medical profession's complicity in the torture of defenceless women. These shocking revelations generated considerable public concern and raised questions regarding the proper role of doctors in relation to the state, the rights of the incarcerated, and the scope of individual bodily autonomy. Ultimately, the suffrage campaign turned the horror of force-feeding to their political advantage. Not only did it lend sympathy to the cause, the harrowing experience of being forced to eat, drink and digest against one's will helped to forge solidarity among protesters.

At a meeting at St James's Hall in July of 1909, Wallace Dunlop addressed her fellow WSPU members, many of whom 'mounted their seats, handkerchiefs and scarves fluttering' to hear what the pioneering hunger striker had to say. Her speech, as the campaign newspaper *Votes for Women* recorded, praised the 'heroines' that had come in her wake, who had been forced to endure the gnawing in their guts while confined to 'dark, lonely cells, deprived of everything that would have ameliorated their sufferings', but whose torments had not been in vain. On release from prison, hunger strikers were awarded military-style medals to commemorate their courage and admirable commitment to the cause.[307] Often these were distributed at celebratory breakfasts, which in London were usually hosted by the Eustace Miles Restaurant, an establishment that was known to be friendly to the cause. In 1908 an article in *The Times* relayed:

Forty woman suffragists who had completed their term of fourteen days' imprisonment, to which they were sentenced at Westminster Police Court on March 21st, in default of payment of fines, for disorderly conduct outside the Houses of Parliament on the previous day, were released from Holloway Jail yesterday morning. Twelve others had been released before, having had their fines paid against their will or in special circumstances. Of the nine still remaining in prison, one – Miss King Townsend, of Bradford – will, it is expected, leave the prison on the Monday, and the others in a fortnight's time. The women were met at the prison gates by Mrs Pankhurst, Mrs Cobden-Sanderson, Mrs Edith How-Martyn and other members of the Women's Social and Political Union, and a number of their relatives and friends. Headed by a brass band, which played on the way, they marched in procession to Caledonia Road station where a train was taken for Holborn. There the procession was reformed, and the women marched down Kingsway to the Strand, and thence to the Eustace Miles Restaurant at Charing Cross where the released suffragists were entertained at breakfast.[308]

To march through the centre of London – singing campaign songs and throwing bouquets of flowers to befuddled onlookers – was a marvellous act of defiance that recast the ignominy of incarceration into a memorialisation of the remarkable 'intestinal fortitude' of protesting suffragettes.

Their choice of breakfast venue is telling. In 1912 Eustace Miles found himself in hot water when he was called before the magistrate's court to explain why he was

'harbouring dangerous criminals', as several of the suffragettes who frequented his restaurant were wanted by the
authorities. He refused, rather gallantly, to reveal the identities of his customers and the restaurant continued to serve
as a kind of social club for the WSPU who programmed a
public lecture series in one of its upstairs rooms and set
up campaign pitches on the pavement outside. Meat-free
establishments were important material sites for socialis-
ing, organising and campaigning, because vegetarianism
remained a key moral touchstone for many suffragettes.
Animal rights and women's rights had long been intertwined: early vegetarian reformers appealed to women on
the grounds that they were naturally more sympathetic
to the plight of defenceless creatures and doctors often
argued that women should eat less flesh than men. These
arguments were, however, grounded in an image of woman
as somehow softer than man, more naturally attuned to
the animal world. The radical women of the late nineteenth century flipped this logic on its head: it was not
that women were – like the animals – man's inferior, but
rather that men were slaves to violent impulses and that
their excessive consumption of meat was evidence of their
innate cruelty. In *Shafts*, an early feminist journal established in the 1890s, the writer Edith Ward argued for the
adoption of a meat-free diet on the basis that 'the case for
the animal is the case for women'.[309] According to Ward,
the oppression of women and the exploitation of non-human animals resulted from the same system of power: to
topple the gender hierarchy it would be necessary to also
address inequalities between humans and animals.

It is of some significance that the first meal that many

released suffragettes ate as free women was a meat-free one. Their enjoyment of a fleshless breakfast, in the assurance of the righteousness of their cause and surrounded by allies, placed the question of what to eat at the centre of political change. One of the major victories secured by the Women's Vegetarian Union was to force British prisons to provide meat-free meals to inmates. In the years leading up to the First World War so many suffragettes were refusing to consume flesh that prison authorities eventually agreed to provide vegetarian options.

This small victory is part of the story of how women in Britain won the vote. Which, read alongside the harrowing history of hunger strikes and force-feeding, demonstrates how the politics of digestion have shaped the course of world-changing events. The gut was, throughout the early decades of the twentieth century, a rich ferment of gender ideologies that helped to define the limits of conventional masculinity and femininity. For men in this period the stomach was both an organ that had to be disciplined, trained to meet the demands of modern life, and the site of a kind of untrained, instinctual masculinity that risked being blunted by the softening influence of civilisation. In contrast, the supposed weakness of women's guts was consistently cited as evidence of constitutional fragility and used to justify their exclusion from public life. However, while products like Bile Beans insisted – with the support of medical authorities – that the vulnerability of the female digestive system rendered them incapable of doing anything more than homemaking and childbearing, the courageous actions of imprisoned suffragettes defied expectation and recast the stomach as a radical political tool. Refusing the association of gut health with reproductive capacity or conventional beauty standards, these pioneering 'New Women' instead insisted on digestion as a strategy for bringing about change and winning new freedoms.

Listen to Your Gut

In the long history of the gut only once has the organ been imagined to be responsible for the bankruptcy of a company. Published in 1940, Graham Greene's short story 'Alas, Poor Maling' tells the tale of a rather unfortunate individual, happily employed by a large commercial printer but plagued by a mysterious stomach complaint. He has a strange case of 'borborygmi', tummy rumbles that usually signify 'quite harmless indigestion'. But not so for the eponymous Maling, who complains to the narrator that his stomach has 'an ear' and can mimic sounds from the world around it. One night, after a dinner at the 'Piccadilly Hotel in honour of a party of provincial printers', his gut begins to play the opening bars of a Brahms concerto, interrupting the party atmosphere with its 'sad, plangent' timbre. No accurate diagnosis can be made because in the doctor's surgery his stomach has 'always lain quiet' and now, Maling complains, 'I can't enjoy food anymore. I never know what's going to happen afterwards.' The situation comes to a head during an important company meeting, during which an urgently needed merger is due to be brokered. But Maling's parrot-like belly begins to sound an air-raid siren, sending everyone running for shelter and scuppering the deal. Too embarrassed to admit that the noise is emitting from his gut, whose rumbles have been provoked by a brief mention of pudding, Maling hopes that 'for once the stomach would do the right thing and make amends'. His belly has not, however, learned how to sound the 'all clear' and he remains huddled underground with his terrified bosses as the company's finances fall apart.[310] The vociferous belly, full of copycatting gurgles and sly mocking groans, brings the business, with its distinguished 'offices in Fetter Lane', to its knees.

As was the case with the young woman plagued by borborygmus and treated by Erasmus Darwin in 1794, the boisterous gut presents a problem in Greene's amusing short story because its interjections into the social realm directly contravene the rules that govern the world. Had Maling admitted that his stomach was the source of the noise, then financial disaster could have easily been averted, but to do so would have been to drag the messy, shame-inducing work of the viscera into the light. For Maling, it would have involved admitting the reality of his body, one that consumes, digests and defecates, to employers who value only his brain and specifically its knowledge of the intricacies of income tax law. The doctors he consults attribute his gastric difficulties to poor digestion, but Greene leaves open other possibilities regarding the source of his character's stomach rumbles. Like the opinionated belly dramatised by Sydney Whiting's *Memoirs of a Stomach*, his gut is a truth-teller intent on exposing the grubby hypocrisies and grasping professional aspirations of its owner. Whether possessed by an outside force, animated by some unknown pathology or an expression of deeply buried impulses in the unconscious, Maling's noisy stomach is a mischievous, rebellious force that lays bare folly and punctures pretension.

Away from literature, modern medicine is increasingly interested in what the rumbles of the gut might reveal. These bodily reverberations are part of the normal digestive process, but they can sometimes speak to the existence of underlying health problems. An excessively vocal stomach can be a symptom of gastric issues like colitis, coeliac disease or diverticulitis, conditions that are on the rise in countries around the world. It is estimated, for example, that 10–20 per cent of people in Britain suffer from irritable bowel syndrome (IBS), a chronic condition that plagues the large intestine and causes a whole host of uncomfortable symptoms including cramping, abdominal pain, diarrhoea and constipation.[311] IBS can be managed but it is notoriously difficult to diagnose as it presents differently from one person to the next, just as its effects tend to vary in severity from one day to the next. Recently, a team of Australian scientists have attempted to address this issue by tuning into the audio of the digestive system. The Noisy Guts Project has developed a special acoustic belt that, once strapped firmly around the patient's abdomen, records the gassy burbles of the gut, which are then analysed alongside data on heart rate and skin temperature to determine a diagnosis.[312] Avoiding invasive procedures like the colonoscopy, the belt allows doctors to map the soundscape of the belly and to listen out for any discordant notes that might signal a disturbance deep in the viscera.

Rumbles has been similarly interested in what the gut

might have to say for itself. Invested in the gut's power – as a part of the body, as a set of actions and as a metaphor – to disrupt, unsettle and upend, it has tried to better understand how this incredible organ speaks to the world. As its history suggests, the gut has a lively cultural and linguistic life that implicates it in complicated debates around subjectivity, technology, sexuality, spirituality, nationhood and identity. These entanglements point to a fundamental connection between mind and belly, one that exceeds the limitations of modern biomedicine and reveals the remarkable degree to which our embodied experiences are shaped by the worlds that we inhabit. Intelligence has long been viewed as primarily a property of the brain, but many of the stories uncovered by this book suggest that it might be more multiple, more dispersed, and more bodily than we currently acknowledge. Following Stanley Cavell's revelation that to be human is to be 'strung out on both sides of a belly', it is clear that digestive processes – consumption, absorption, defecation – play an important role in how we shape and come to understand the self.

Yet as with the unhappy protagonist of 'Alas, Poor Maling', whose troublesome stomach is forever overruling his express wishes, the gut is also a site at which the self can feel under threat. There is a great deal of anxiety associated with the work of digestion, apparent in both the vast sewer systems we have constructed to distance ourselves from its waste products and in the taboos that attach to it so readily. Eating entails an intensely intimate encounter with the world that reveals the boundary between exterior and interior to be a profoundly permeable one. Today this ontological wobble finds expression in everything from worries about over-processed foods to rising levels of gastric illness, and history – as we have seen – documents the stomach as a site of death, disease, melancholic moods and demonic possession. What this history also offers up, however, are other ways of imagining, speaking and listening to the gut that may help us to forge a friendlier relationship with it. Recent scientific research has already encouraged us to see the gut in a fresh light, as an organ of infinite complexity, containing multitudes of mysterious microbial life, deeply networked into the rest of the body, and strangely entwined with the intricacies of our psychic lives, but history can broaden our lexicon of the belly and provide us with new (old) ways to think about its processes. We have inherited a richly storied gut, full of oddities, intrigue and tales that feel both familiar and strange. If you listen carefully to its rumbles, you might find that the gut has more to say about the world than you thought.

Acknowledgements

Rumbles started in the pandemic and was written – slowly and across a shifting confederacy of beds, desks and kitchen tables – over the first few years of my son's life. My deepest thanks go to his other mother Laura Guy, whose support, encouragement and insight made it all possible.

Slow to digest (sorry), *Rumbles* is the result of countless conversations with friends, family and colleagues. Thanks go to the Max Planck Center for the History of Emotions for hosting me and allowing me to present an early version of the manuscript; to Tamar Novick and Maria Pirogovskaya who welcomed me so warmly to the Max Planck Institute for the History of Science and made me think more carefully about where waste fits into the story of the gut; and to Lucy Beech for her intellectual generosity and collaborative spirit. Some of my most sustained discussions have been with those involved with the Scottish Gut Project; particular thanks go to Manon Mathias, Kirsty Hendry and Maya Hey for always having another story about the belly to tell me.

This book originated in conversations with Kirty Topiwala and Cecily Gayford, who supported and helped to shape the project in its early days. I am endlessly grateful to Martin Redford, my first ever literary agent, for taking me out for fancy drinks in Soho and convincing me that I could write a book that people might read. *Rumbles* has been beaten into shape by my fantastic Wellcome Collection editors Ellen Johl and Fran Fabriczki, by Georgina Difford at Profile Books, by Patrick Taylor and Lorraine Green; I appreciate all the time that you have spent on *Rumbles* – your knowledge, curiosity and expertise have made it a much better book.

Finally, thank you to my wonderful friends, whose willingness to discuss all matters of the gut – no matter how intimate – is truly an inspiration.

Notes

1   Erasmus Darwin, *Zoonomia; or, The Laws of Organic Life* (London: J. Johnson, 1794), p. 104

2   Ibid.

3   Michael C. Schoenfeldt, *Bodies and Selves in Early Modern England: Physiology and Inwardness in Spenser, Shakespeare, Herbert, and Milton* (Cambridge: Cambridge University Press, 1999), p. 25

4   François Rabelais, *Gargantua and Pantagruel* (1564), Book IV, ch. 58, *Oeuvres Complètes*, ed. Jacques Boulenger (Paris: Pléiade, 1953), pp. 697–8

5   Sydney Whiting, *Memoirs of a Stomach: Written by Himself, That All Who Eat May Read* (London: W. E. Painter, 1853), p. 105

6   Michael D. Gershon, *The Second Brain: A Groundbreaking New Understanding of Nervous Disorders of the Stomach and Intestine* (New York: HarperCollins, 1998), p. 84

7   John F. Cryan, Kenneth J. O'Riordan, Caitlin S. M. Cowan et al., 'The Microbiota–Gut–Brain Axis', *Physiological Reviews* 1:99(4) (2019), 1877–2013

8   'gut, n.' *OED Online*, Oxford University Press, oed.com/dictionary/gut_n

9   Sudabeh Alatab, Sadaf G. Sepanlou, Kevin Ikuta et al., 'The Global, Regional, and National Burden of Inflammatory Bowel Disease in 195 Countries and Territories, 1990–2017: A Systematic Analysis for the Global Burden of Disease Study 2017', *Lancet Gastroenterology & Hepatology* 5:1 (2020), 17–30

10   'Europe Is "Failing" to Deal with Chronic Digestive Disease
     Burden', United European Gastroenterology, 14 May 2018
11   'history, n.' OED Online, Oxford University Press, oed.com/
     dictionary/history_n
12   Mark Neocleous, The Politics of Immunity: Security and the Policing
     of Bodies (London: Verso, 2022), p. 7
13   Raymond Williams, 'Culture Is Ordinary', in Resources of Hope:
     Culture, Democracy, Socialism (London: Verso, 1989), p. 4
14   Clair R. Martin, Vadim Osadchiy, Amir Kalani and Emeran
     A. Mayer, 'The Brain–Gut–Microbiome Axis', Cellular and
     Molecular Gastroenterology and Hepatology 6:2 (2018), 133–48;
     Aitak Farzi, Esther E. Fröhlich and Peter Holzer, 'Gut Microbiota
     and the Neuroendocrine System', Neurotherapeutics 15:1 (2018),
     5–22; T. G. Dinan and J. F. Cryan, 'Melancholic Microbes: A Link
     Between Gut Microbiota and Depression?', Neurogastroenterology
     & Motility 25:9 (2013), 713–19; Jeffrey M. Lackner, Brian M.
     Quigley and Edward B. Blanchard, 'Depression and Abdominal
     Pain in IBS Patients: The Mediating Role of Catastrophizing',
     Psychosomatic Medicine 66:3 (2004), 435–41
15   Jan Plamper, The History of Emotions: An Introduction (Oxford:
     Oxford University Press, 2015); Barbara H. Rosenwein, 'Worrying
     About Emotions in History', American Historical Review 107:3
     (2002), 821–45
16   Frederick Kaufman, A Short History of the American Stomach
     (Boston and New York: Mariner Books, 2008), p. xiv
17   Corinna Treitel, Eating Nature in Modern Germany: Food,
     Agriculture and Environment, c.1870 to 2000 (Cambridge:
     Cambridge University Press, 2017), p. 95
18   Stanley Cavell, Disowning Knowledge: In Six Plays of Shakespeare
     (Cambridge: Cambridge University Press, 1987), p. 176
19   George Cheyne, Dr Cheyne's Account of Himself and of His
     Writings (London: J. Wilford, 1744), p. 2
20   George Cheyne, The English Malady; or, A Treatise of Nervous
     Diseases of All Kinds, as Spleen, Vapours, Lowness of Spirits,
     Hypochondriacal, and Hysterical Distempers, &c. (London: G.
     Strahan, 1733), p. 361
21   Ibid, p. 49
22   E. M. Tansey, 'Sir William Maddock Bayliss (1860–1924)';

J. Barcroft and Anita McConnell, 'Ernest Henry Starling (1866–1927)', *Oxford Dictionary of National Biography*

23 Ernest H. Starling, 'The Croonian Lectures. I. On the Chemical Correlation of the Functions of the Body', *Lancet* 166 (1905), 339–41

24 William Bayliss and Ernest H. Starling, 'The Movements and Innervation of the Small Intestine', *Journal of Physiology* 24:2 (1899), 99–143 (p. 99)

25 C. Sherrington and Caroline Overy, 'John Newport Langley (1852–1925)', *Oxford Dictionary of National Biography*

26 John Newport Langley, 'The Sympathetic and Related Systems of Nerves', in *Schäfer's Textbook of Physiology*, ed. E. A. Sharpey-Schäfer (Edinburgh & London: Young J. Pentland, 1900), pp. 593–616

27 Byron Robinson, *The Abdominal and Pelvic Brain: With Automatic Visceral Ganglia* (Hammond, IN: Frank S. Betz, 1907), p. 124

28 Gershon, *The Second Brain*, p. 17

29 Ibid.

30 Vladimir Grubišić and Brian Gulbransen, 'Enteric Glia: The Most Alimentary of All Glia', *Journal of Physiology* 15:595(2) (2017), 557–70; Michael Schemann, Thomas Frieling and Paul Enck, 'To Learn, to Remember, to Forget: How Smart Is the Gut?', *Acta Physiologica* 228:1 (2020)

31 Nick J. Spencer, Lee Travis, Lukasz Wiklendt et al., 'Long Range Synchronization Within the Enteric Nervous System Underlies Propulsion Along the Large Intestine in Mice', *Communications Biology* 4 (2021)

32 René Descartes, *Discourse on the Method of Rightly Conducting the Reason and Seeking Truth in the Sciences* (1637), pp. 19–20

33 René Descartes, *Treatise on Man* (1662)

34 René Descartes, *Passions of the Soul* [1649], trans. Stephen Voss (Indianapolis: Hackett Publishing, 1989)

35 Ibid., p. 36

36 See Lisa Shapiro, 'Descartes's Pineal Gland Reconsidered', *Midwest Studies in Philosophy* 35 (2011), 259–86

37 Galen, *On the Usefulness of the Parts of the Body (De usu partium)*, trans. Margaret Tallmadge May, 2 vols (Ithaca, NY: Cornell University Press, 1968), Vol. 1, p. 205

38 Thomas Willis, *Cerebri anatome: cui accessit nervorum descriptio et usus* (London, 1664)

39 Nicolas Fontaine, *Mémoires pour servir a l'histoire de Port-Royal* (1738)

40 Erica Fudge, *Brutal Reasoning: Animals, Rationality, and Humanity in Early Modern England* (Ithaca, NY: Cornell University Press, 2019), p. 180

41 Roger Fulford (ed.), *Darling Child: Private Correspondence of Queen Victoria and the German Crown Princess of Russia, 1871–1878* (London: Evans Brothers, 1976), p. 185

42 Royal Commission on Vivisection, *Report of the Royal Commission on the Practice of Subjecting Live Animals to Experiments for Scientific Purposes; With Minutes of Evidence and Appendix* (London: George Edward Eyre and William Spottiswoode, 1876), p. 183

43 Ibid., pp. 138, 272

44 See N. Katherine Hayles, 'Microbiomimesis: Bacteria, Our Cognitive Collaborators', *Critical Inquiry* 47:4 (Summer 2021)

45 Elizabeth M. Knowles, 'Blue Devils', *Oxford Dictionary of Phrase and Fable* (Oxford: Oxford University Press, 2000)

46 John Keats to John Reynolds, 18 September 1819, in *The Letters of John Keats: Vol. 2, 1819–1821*, ed. Hyder Edward Rollins (Oxford: Oxford University Press, 2000), pp. 166–8; Byron, *Don Juan* (London: John Murray, 1819)

47 James Johnson, *Essay on Indigestion; or, Morbid Sensibility of the Stomach* [1827] (London: S. Highley, 1840), p. 30

48 Alessandro Benedetti, 'History of the Human Body' [1497], in *Studies in Pre-Vesalian Anatomy: Biography, Translations, Documents*, ed. L. R. Lind (Philadelphia: American Philosophical Society, 1975), p. 90

49 N. Henessey, 'Dietetics in Relation to Mental Culture', *The Dietetic Reformer and Vegetarian Messenger* (1 February 1873)

50 Steven Shapin, '"You Are What You Eat": Historical Changes in Ideas About Food and Identity', *Historical Research* 87:237 (2014), 377–92 (p. 378)

51 Larry Duffy, 'Textual (In)Digestions in Flaubert, Zola and Huysmans: Accumulation, Extraction, Regulation', in *Gut Feeling and Digestive Health in Nineteenth-Century Literature, History*

and Culture, ed. Manon Mathias and Alison M. Moore (New York: Palgrave, 2018), p. 181; Noah Webster, *Webster's Collegiate Dictionary*, 5th edn (New York: G. & C. Merriam, 1941); W. C. Orr, 'Sleep and Functional Bowel Disorders: Can Bad Bowels Cause Bad Dreams?', *American Journal of Gastroenterology* 95:5 (2000), 1118–21

52 Carole Levin, *Dreaming the English Renaissance: Politics and Desire in Court and Culture* (New York: Palgrave Macmillan, 2008), p. 62

53 Sasha Handley, *Sleep in Early Modern England* (New Haven and London: Yale University Press, 2016), pp. 22–3

54 Robert Macnish, *The Philosophy of Sleep* (Glasgow: W. R. M'Phun, 1827), p. 53

55 Sigmund Freud, *The Interpretation of Dreams* [1899], ed. and trans. James Strachey (New York: Basic Books, 1955), p. 483

56 Ibid., p. 130

57 cheesedeutung.tumblr.com

58 Katherine Roeder, *Wide Awake in Slumberland: Fantasy, Mass Culture, and Modernism in the Art of Winsor McCay* (Jackson: University of Mississippi Press, 2014), pp. 155–6

59 Ibid., p. 14

60 Charles Dickens, *A Christmas Carol* [1843] (Philadelphia: J. B. Lippincott, 1915), p. 24

61 Shane McCorristine, *Spectres of the Self: Thinking About Ghosts and Ghost-Seeing in England, 1750–1920* (Cambridge: Cambridge University Press, 2010), pp. 27, 31

62 Reginald Scot, *The Discoverie of Witchcraft* (London: William Brome, 1584)

63 Caroline Oates, 'Cheese Gives You Nightmares: Old Hags and Heartburn', *Folklore* 114:2 (2003), 205–25 (p. 221)

64 Porphyry, *On Abstinence from Killing Animals*, trans. Gillian Clark (London: Bloomsbury, 2000), p. 73

65 Christopher Kissane, *Food, Religion and Communities in Early Modern Europe* (London: Bloomsbury, 2018), p. 108

66 In his article 'Historians as Demonologists: The Myth of the Midwife-Witch', *Social History of Medicine* 3:1 (1990), 1–26, David Harley characterised the claim made by several twentieth-century historians that midwives working in the sixteenth and

seventeenth centuries were frequently prosecuted for witchcraft as grossly exaggerated. In his research he uncovered only one English trial in which a midwife was accused, leading Harley to claim that historians had been engaged in the creation of a 'powerful myth'. These claims have since been refuted by, among others, Barbara Ehrenreich and Deirdre English, who counter that 'the association that witch hunters made between witches and midwives in Europe is inescapable' and cite archival research that documents several examples from across Germany in which midwives were indeed prosecuted as witches (see their *Witches, Midwives, and Nurses: A History of Women Healers*, 2nd edn (New York: Feminist Press, 2010)).

67 Heinrich Kramer, *The Malleus Maleficarum* [1487], trans. Montague Summers (London: John Rodker, 1928), p. 66

68 Nancy Caciola, *Discerning Spirits: Divine and Demonic Possession in the Middle Ages* (Ithaca, NY: Cornell University Press, 2003), p. 171

69 Boyd Brogan, 'His Belly, Her Seed: Gender and Medicine in Early Modern Demonic Possession', *Representations* 147:1 (Summer 2019), 1–25 (p. 7)

70 Nicholas Rémy, quoted in Brogan, p. 7

71 Mary Post, *Etiquette in Society, in Business, in Politics, and at Home* (New York and London: Funk & Wagnalls, 1922) pp. 177–230

72 Peggy Post, Anna Post, Lizzie Post and Daniel Post Senning, *Emily Post's Etiquette, 18th edn: Manners for a New World* (New York: William Morrow, 2011)

73 Isabella Beeton, *Book of Household Management* (London: S. O. Beeton, 1861), p. 905

74 Ibid.

75 Norbert Elias, *The Civilizing Process: Sociogenetic and Psychogenetic Investigations* [1939], rev. edn, ed. Eric Dunning, Johan Goudsblom and Stephen Mennell, trans. Edmund Jephcott (Oxford: Blackwell, 2000)

76 Charles Vyse, *The New London Spelling Book* (London: G. Robinson, 1778), quoted in Keith Thomas, *In Pursuit of Civility: Manners and Civilization in Early Modern England* (New Haven and London: Yale University Press, 2018), p. 51

77 Ken Albala, *Eating Right in the Renaissance* (Berkeley: University of California Press, 2002), p. 218

78 Humphrey Brooke, *A Conservatory of Health, Comprised in a Plain and Practical Discourse upon the Six Particulars Necessary for Man's Life* (London: 1650), quoted in Albala, *Eating Right in the Renaissance*, p. 218

79 Plato, *Phaedrus* 238a–b, trans. Harold N. Fowler, Loeb Classical Library 36 (Cambridge, MA: Harvard University Press, 1982), p. 447

80 See Barbara H. Rosenwein, *Emotional Communities in the Middle Ages* (Ithaca, NY: Cornell University Press, 2006), p. 48

81 Thomas Aquinas, *Summa Theologica* ii-ii. Q148.A4

82 William Vaughan, *Naturall and Artificial Directions for Health: Deriued from the Best Philosophers, as Well Moderne, as Auncient* (London: Richard Bradocke, 1600)

83 'An Homilie Against Gluttony and Drunkennesse', quoted in Schoenfeldt, *Bodies and Selves in Early Modern England*, p. 24

84 'The Publisher to the Reader', in Roger Crab, *The English Hermite; or, Wonder of this Age* (London: 1655), p. 1

85 See Kerry S. Walters and Lisa Portmess, *Religious Vegetarianism: From Hesiod to the Dalai Lama* (New York: State University of New York Press, 2001) for a detailed history of meat-avoidance and religious tradition.

86 Ian Miller, 'The Gut–Brain Axis: Historical Reflections', *Microbial Ecology in Health and Disease* (2018) 29:2, 1–9 (p. 2)

87 Thomas Cogan, *The Haven of Health: Chiefly Gathered for the Comfort of Students* (London: William Norton, 1584)

88 Ibid., pp. 69, 71, 111, 102

89 Ibid., pp. 129, 128, 146, 132

90 John Selden, 'Preface', *Titles of Honor* (London: John Helme, 1614)

91 Joseph Firth, James E. Gangwisch, Alessandra Borsini, Robyn E. Wootton and Emeran A. Mayer, 'Food and Mood: How Do Diet and Nutrition Affect Mental Wellbeing?', *British Medical Journal* 369 (2020)

92 Cogan, *The Haven of Health*, pp. 129–30

93 Jan Purnis, 'The Stomach and Early Modern Emotion', *University of Toronto Quarterly* 79:2 (Spring 2010), 800–18 (p. 803)

94   Joseph Duchesne, *Le Pourtraict de la santé* (Paris: Claude Morel, 1606)

95   Diocletian Lewis, *Our Digestion; or, My Jolly Friend's Secret* (Philadelphia and Boston: G. Maclean, 1872), p. 101

96   Mikhail Bakhtin, *Rabelais and His World* [1965], trans. Hélène Iswolsky (Bloomington: Indiana University Press, 1984), p. 10

97   Gurdon Saltonstall Hubbard quoted in William Beaumont, *Experiments and Observations on the Gastric Juice, and the Physiology of Digestion* (Plattsburgh: F. P. Allen, 1833), p. 10

98   Jesse S. Myer, *Life and Letters of Dr. William Beaumont* (St. Louis: C. V. Mosby, 1912), pp. 107–8

99   René Antoine Ferchault de Réaumur quoted in Henry Smith Williams, *A History of Science*, Vol. 4 (Frankfurt: Outlook Verlag, 2018), p. 54

100 Alexander W. Blyth, 'Diet in Relation to Health and Work', in *Health Exhibition Literature*, Vol. 4 (William Clowes and Sons: London, 1884), pp. 251–354

101 Vaughan, *Naturall and Artificiall Directions for Health*, p. 168

102 Andreas Vesalius quoted in Schoenfeldt, *Bodies and Selves in Early Modern England*, p. 26

103 Edmund Spenser, *The Faerie Queene*, ed. C. O'Donnell and Thomas Roche (London: Penguin, 2003), II. 30

104 John Abernethy, 'An Enquiry into the Probability and Rationality of Mr Hunter's Theory of Life', in *Physiological Lectures, Addressed to the College of Surgeons* (London: Longman, Hurst, Rees, Orme, Brown and Green, 1825), p. 53

105 Marilyn Butler, '*Frankenstein* and Radical Science', in *Frankenstein*, ed. J. Paul Hunter (New York: Norton, 1996), pp. 302–13

106 Ibid., p. 307

107 Lawrence was alluding to a line from Alexander Pope's poem 'An Essay on Man' (1733–4).

108 Abernethy, 'An Enquiry into the Probability and Rationality of Mr Hunter's Theory of Life', p. 48

109 John Hunter, 'On the Digestion of the Stomach after Death', *Proceedings of the Royal Society of London* (1772), 447–54 (p. 453)

110 William Hunter, *Introductory Lectures* (London: J. Johnson, 1784), p. 67

111 See Ruth Richardson, *Death, Dissection and the Destitute* (New York: Routledge and Kegan Paul, 1987) for a full account of the body-snatching scandal.

112 Quoted in William Osler, 'William Beaumont: A Pioneer American Physiologist', in *Experiments and Observations on the Gastric Juice and the Physiology of Digestion* [1833] (Mineola: Dover Publications, 1996), p. xiii

113 Mary Roach, *Gulp: Adventures on the Alimentary Canal* (New York: W. W. Norton, 2013), p. 98

114 Beaumont, *Experiments and Observations on the Gastric Juice, and the Physiology of Digestion* (1833), p. 90

115 Roach, *Gulp*, pp. 100–1

116 Andrew Combe, 'Preface', in William Beaumont, *Experiments and Observations on the Gastric Juice, and the Physiology of Digestion* (Edinburgh: Maclachlan and Stewart, 1838), pp. xiv–xv

117 Walter B. Cannon and George Higginson, 'The Book of William Beaumont after One Hundred Years', *Bulletin of the New York Academy of Medicine* 9:10 (1933), 568–84 (p. 583)

118 Walter B. Cannon, *Bodily Changes in Pain, Hunger, Fear and Rage: An Account of Recent Researches into the Function of Emotional Excitement* (New York: D. Appleton and Co., 1915)

119 Cannon and Higginson, 'The Book of William Beaumont after One Hundred Years', p. 581

120 Michail Mantzios, '(Re)Defining Mindful Eating into Mindful Eating Behaviour to Advance Scientific Enquiry', *Nutrition and Health* 27:4 (2021), 367–71

121 Robert Burton, *The Anatomy of Melancholy*, 3 vols [1621] (London: William Tegg, 1863), p. 4

122 Ibid., p. 551

123 Ibid., p. 72

124 Ibid., p. 200

125 Robert Whytt, *Observations on the Nature, Causes and Cure of Those Disorders which Have Been Commonly Called Nervous, Hypochondriac or Hysteric* (Edinburgh: T. Becket, 1765), p. 127

126 Cheyne, *The English Malady*, p. 44

127 Ibid., p. 52

128 Ibid., p. 54

129 Alexander Crichton, *An Inquiry into the Nature and Origin of*

Mental Derangement, Vol. 2 (London: T. Cadell & W. Davies, 1798), p. 30

130 Richard Blackmore, *A Treatise of the Spleen and Vapours* (London: J. Pemberton, 1726), p. 91

131 Cogan, *The Haven of Health*, pp. 116–17

132 Anne-Charles Lorry, *Essai sur le alimens* (Paris: Vincent, 1754), p. 236

133 Samuel Auguste André David Tissot, *An Essay on Diseases Incident to Literary and Sedentary Persons*, trans. J. Kirkpatrick (London: J. Nourse, 1769), pp. 111, 112

134 Samuel Johnson to James Boswell, 2 July 1776, in *The Life of Samuel Johnson*, Vol. 3 [1791] (London: Office of the National Illustrated Library, 1851), p. 64

135 Samuel Taylor Coleridge to George Beaumont, 1 February 1804, in *Memorials of Coleorton: Being Letters from Coleridge, Wordsworth and His Sister, Southey, and Sir Walter Scott to Sir George and Lady Beaumont, 1803–1834*, Vol. 1 (Boston: Houghton, Mifflin and Co., 1887), p. 43

136 Thomas Trotter, *A View of the Nervous Temperament* (Newcastle: Longman, Hurst, Rees and Orme, 1807), p. 45

137 Roy and Dorothy Porter, *In Sickness and in Health: The British Experience, 1650–1850* (London: Fourth Estate, 1988), pp. 203–10

138 James Kennaway and Jonathan Andrews, 'The Grand Organ of Sympathy: Fashionable Stomach Complaints and the Mind in Britain, 1750–1850', *Social History of Medicine* 32:1 (2019), 57–79 (p. 69)

139 Elizabeth L. Swann, *Taste and Knowledge in Early Modern England* (Cambridge: Cambridge University Press, 2020), p. 12

140 William Shakespeare, *Love's Labour's Lost*, ed. G. R. Hibbard (Oxford: Oxford University Press, 2008), iv. 2, 25–30

141 Spenser, *The Faerie Queene*, II. 20

142 Ben Jonson, *Timber; or, Discoveries Made Upon Men and Matter* (London, 1641), p. 127

143 Denise Gigante, *Taste: A Literary History* (New Haven and London: Yale University Press, 2005), p. 16

144 Francis Bacon, 'Of Studies' [1597], *The Major Works*, ed. Brian Vickers (Oxford: Oxford University Press, 2002), p. 81

145 Desiderius Erasmus, *Ciceronianus; or, a Dialogue on the Best Style*

of Speaking [1528] (Virginia: University of Virginia Press, 1972), p. 4

146 Seneca, *Epistles*, quoted in Swann, p. 43

147 John Milton, *Areopagitica: A Speech of Mr John Milton for the Liberty of Unlicenc'd Printing, to the Parlament of England* (1644), in *Censorship and the Press, 1580–1720, Vol. 2: 1640–1660*, eds. Geoff Kemp and Jason McElligott (London: Pickering & Chatto, 2009), pp. 95–215

148 Ibid., p. 154

149 Ibid.

150 Queen Elizabeth I quoted in Helen Smith, '"More swete vnto the eare/than holsome for ye mynde": Embodying Early Modern Women's Reading', *Huntington Library Quarterly* 73:3 (2010), 413–32

151 Edward Reynolds, *A Treatise of the Passions and Faculties of the Soul of Man* (London: Robert Bostock, 1640), p. 164

152 Anne C. Vila, 'The *Philosophe*'s Stomach: Hedonism, Hypochondria, and the Intellectual in Enlightenment France', in *Cultures of the Abdomen: Diet, Digestion, and Fat in the Modern World*, eds Christopher E. Forth and Ana Carden-Coyne (New York: Palgrave, 2005), pp. 89–104 (p. 89)

153 Samuel Taylor Coleridge, *Philosophical Lectures* (1819), quoted in Gigante, *Taste*, p. 13

154 Lucy McDonald, 'The Sandwich That Changed Lunch Forever', *Daily Mail*, 23 April 2010

155 'The Work Issue', *New York Times Magazine*, 28 February 2016

156 Nupur Amarnath, 'Can I Eat My Office Lunch at My Desk?', *The Times of India*, 16 October 2016; Zaria Gorvett, 'The Norwegian Art of the Packed Lunch', BBC News, 3 January 2019; Suthentira Govender, 'Work Through Lunch? You're Giving Your Company Over R500k', *Sunday Times* (South Africa), 28 October 2018

157 Roger Cohen, 'France's Latest Covid Measure: Letting Workers Eat at Their Desks', *New York Times*, 22 February 2021

158 Andrea Broomfield, *Food and Cooking in Victorian England: A History* (Westport, CT: Praeger, 2007), p. 24

159 etymonline.com/word/lunch

160 E. P. Thompson, 'Time, Work-Discipline, and Industrial Capitalism', *Past and Present* 38 (1967), 56–97 (p. 86)

161 'Luncheon Bars', *London Society* 17:99 (March 1870), 241

162 Alfred Haviland, *Hurried to Death; or, a Few Words of Advice on the Danger of Hurry and Excitement, Especially Addressed to Railway Travellers* (London: Henry Renshaw, 1868), p. 13

163 'Diet and Dyspepsia', *All the Year Round*, 6 February 1886, 545

164 W. M. Wallace, *A Treatise on Desk Diseases: Containing the Best Methods of Treating the Various Disorders Attendant Upon Sedentary and Studious Habits* (London: T. Griffiths, 1826), p. 7

165 Broomfield, *Food and Cooking in Victorian England*, p. 54

166 C. P. Newcombe, *The Manifesto of Vegetarianism* (London: Vegetarian Society, 1911), p. 13

167 George Gissing, *New Grub Street* [1891], ed. John Goode (Oxford: Oxford University Press, 2008), p. 92

168 Michael Heller, *London Clerical Workers, 1880–1914: Development of the Labour Market* (London: Pickering & Chatto, 2011), p. 1

169 Robert Bell, 'On the Treatment of a Stomach', *The Butterfly*, September 1899, 135

170 'Another City Clerk', *The Vegetarian*, 2 February 1889, 67

171 'Rise of the Vegetarian Restaurant', *The Vegetarian*, 5 March 1887, 56

172 Rhodri Hayward, 'Busman's Stomach and the Embodiment of Modernity', *Contemporary British History* 31:1 (2017), 1–23

173 'Alleged Prevalence of Nervous Complaints Considered by Select Committee on the Post Office', *The Postal Clerks' Herald*, 26 March 1910, 576

174 Sue Zemka, *Time and the Moment in Victorian Literature and Society* (Cambridge: Cambridge University Press, 2011), p. 75

175 Vicky Long, 'Situating the Factory Canteen in Discourses of Health and Industrial Work in Britain (1914–1939)', *Le Mouvement social* 2:247 (2014), 65–83 (p. 70)

176 Food (War) Committee meeting, 24 April 1917, cited in Long, p. 71

177 John B. Watson, *Psychological Care of Infant and Child* (London: George Allen and Unwin, 1928), p. 73

178 Ibid., p. 103

179 Jean Walton, *Dissident Gut: Technologies of Regularity, Politics of Revolt* (Edinburgh: Edinburgh University Press, 2024), pp. 93–4

180 Lord Stanley quoted in Lawrence Goldman, *Science, Reform, and*

Politics in Victorian Britain: The Social Science Association 1857–1886 (Cambridge: Cambridge University Press, 2002), p. 198

181 Michael Faraday, 'Letter to the Editor', *The Times*, 7 July 1855

182 Benjamin Disraeli, House of Commons Debate, 15 July 1858, Hansard Vol. 151, cc1508–40

183 *London Gazette*, 21 October 1831

184 Henry Mayhew, *London Labour and the London Poor: A Cyclopaedia of the Condition and Earnings of Those That Will Work, Those That Cannot Work and Those That Will Not Work*, Vol. 2 (London: Griffon, Bohn, 1861), p. 430

185 Ibid., p. 436

186 David S. Barnes, *The Great Stink of Paris and the Nineteenth-Century Struggle against Filth and Germs* (Baltimore: Johns Hopkins University Press, 2006)

187 Quoted in David S. Barnes, 'Scents and Sensibilities: Disgust and the Meaning of Odors in Late Nineteenth-Century Paris', *Historical Reflections* 28:1 (Spring 2002), 21–49

188 Alain Corbin, *The Foul and the Fragrant: Odor and the French Social Imagination* [1982], trans. M. L. Kochan, R. Porter and C. Prendergast (Cambridge, MA: Harvard University Press, 1986), pp. 89–110

189 Barnes, 'Scents and Sensibilities'

190 George Jennings to Commissioners of Sewers for the City of London, 13 December 1858 (London Metropolitan Archives)

191 Quoted in Norbert Elias, 'The Development and Concept of Civilité', in *On Civilization, Power, and Knowledge: Selected Writings*, eds Stephen Mennell and Johan Goudsblom (Chicago: University of Chicago Press, 1997), p. 79

192 Ibid.

193 Johannes Fabian, *Time and the Other: How Anthropology Makes Its Object* (New York: Columbia University Press, 1983), pp. 11–12

194 Alison Moore, 'Kakao and Kaka: Chocolate and the Excretory Imagination of Nineteenth-Century Europe', in *Cultures of the Abdomen*, pp. 51–69

195 Mary Douglas, *Purity and Danger: An Analysis of Concepts of Pollution and Taboo* [1966] (London: Routledge, 2002), p. 143

196 Harold Farnsworth Gray, 'Sewerage in Ancient and Mediaeval Times', *Sewage Works Journal* 12:5 (1940), 939–46 (p. 945)

197 Douglas, *Purity and Danger*, p. 35
198 William Osler, *The Evolution of Modern Medicine: A Series of Lectures Delivered at Yale University on the Silliman Foundation, in April 1913* (New Haven: Yale University Press, 1921)
199 Fedor Galkin, Polina Mamoshina, Alex Aliper et al., 'Human Gut Microbiome Aging Clock Based on Taxonomic Profiling and Deep Learning', *iScience* 23:6 (2020)
200 Aaro Salosensaari, Ville Laitinen, Aki S. Havulinna et al., 'Taxonomic Signatures of Cause-Specific Mortality Risk in the Human Gut Microbiome', *Nature Communications*, 12:2671 (2021). See also utu.fi/en/news/press-release/researchers-discovered-a-gut-microbiota-profile-that-can-predict-mortality
201 Jan Purnis, *Digestive Tracts: Early Modern Discourses of Digestion* (unpublished PhD thesis: University of Toronto, 2010), p. 17
202 Thomas Blount, *Glossographia* (1656), quoted in Purnis, *Digestive Tracts*, p. 17
203 Thomas Ady, *A Perfect Discovery of Witches* (1661), quoted in Leigh Eric Schmidt, 'From Demon Possession to Magic Show: Ventriloquism, Religion, and the Enlightenment', *Church History* 67:2 (1998), 274–304 (p. 281)
204 John Gregory Bourke, *Scatalogic Rites of All Nations: A Dissertation upon the Employment of Excrementitious Remedial Agents in Religion, Therapeutics, Divination, Witchcraft, Love-Philters, etc., in All Parts of the Globe* (Washington, D.C.: W. H. Lowdermilk & Co., 1891), p. 5
205 Ibid., p. 1
206 Moore, 'Kakao and Kaka', p. 61
207 Stephen Greenblatt, 'Filthy Rites', *Daedalus* 111:3 (1982), 1–16 (p. 2)
208 Heinrich Cornelius Agrippa, *Declamation Attacking the Uncertainty and Vanity of the Sciences and the Arts: An Invective Declamation* [1526] (London: Samuel Speed, 1676), p. 318
209 Schoenfeldt, *Bodies and Selves in Early Modern England*, p. 26
210 Joseph Tate, 'Tamburlaine's Urine', in *Fecal Matters in Early Modern Literature and Art: Studies in Scatology*, eds Jeff Persels and Russell Ganim (Aldershot: Ashgate, 2004), pp. 138–53
211 Gualtherus Bruele, *Praxis Medicinae; or, the Physician's Practise:*

Wherein Are Contained All Inward Diseases from the Head to the Foot (London: J. Norton, 1639), p. 365

212 Hannah Marriott, 'Going through the Motions: The Rise and Rise of Stool-Gazing', *Guardian*, 14 March 2021

213 René Goiffon, *Manuel de coprologie clinique* (Paris: Masson et cie., 1921)

214 Jonathan Sawday, *The Body Emblazoned: Dissection and the Human Body in Renaissance Culture* (London and New York: Routledge, 1995), p. 213

215 *Seed Time: The Quarterly of the Fellowship of the New Life*, quoted in W. H. G. Armytage, *Heavens Below: Utopian Experiments in England, 1560–1960* (London: Routledge and Keagan Paul, 1961), pp. 374–5

216 George Orwell, *The Road to Wigan Pier* [1937] (New York: Harcourt Brace, 1958), p. 248

217 William Arbuthnot Lane, 'The Sewage System of the Human Body', *American Medicine* (1923), 267

218 William Arbuthnot Lane, 'Chronic Intestinal Stasis and Cancer', *British Medical Journal* (1923) 745–7 (p. 747)

219 Frank Crane, 'The Colonic' (1916), quoted in James C. Whorton, *Inner Hygiene: Constipation and the Pursuit of Health in Modern Society* (Oxford: Oxford University Press, 2000), p. 55

220 Emil Kraepelin quoted in Mary de Young, *Encyclopaedia of Asylum Therapeutics, 1750s–1950s* (Jefferson, NC: McFarland, 2015)

221 Xiuxia Yuan, Yulin Kang, Chuanjun Zhuo, Xu-Feng Huang and Xueqin Song, 'The Gut Microbiota Promotes the Pathogenesis of Schizophrenia via Multiple Pathways', *Biochemical and Biophysical Research Communications* 512:2 (2019), 373–80

222 William Arbuthnot Lane, 'The Paramount Importance of Effective Intestinal Drainage in Preventing Ill Health and Disease', *American Medicine* 21 (1926), 689–93 (p. 692)

223 See Manon Mathias, 'Autointoxication and Historical Precursors of the Microbiome–Gut–Brain Axis', *Microbial Ecology in Health and Disease* 29:2 (2018), 36–46

224 Edwin Slosson, *Major Prophets of Today* (Boston: Little, Brown, 1914), p. 175, cited in Harvey Levenstein, *Fear of Food: A History*

of *Why We Worry about What We Eat* (Chicago: University of Chicago Press, 2012), p. 34

225 John Harvey Kellogg, *The Itinerary of a Breakfast* (New York and London: Funk & Wagnalls, 1918), p. 97

226 A. C. Field, 'Vegetarianism Scientifically Considered', *Vegetarian Review* 43 (March 1895), 94

227 Newcombe, *The Manifesto of Vegetarianism*, pp. 3–4

228 Whorton, *Inner Hygiene*, p. 173

229 Levenstein, *Fear of Food*, p. 33

230 Arthur Keith quoted in Whorton, *Inner Hygiene*, p. 74

231 Whorton, *Inner Hygiene*, p. 79

232 Michaeleen Doucleff, 'How Modern Life Depletes Our Gut Microbes', NPR, 21 April 2015

233 See Michel Foucault, *Discipline and Punish: The Birth of the Prison* [1975], trans. Alan Sheridan (New York: Pantheon, 1977)

234 'Mr Vaucanson's Letter to the Abbé De Fontaine', in *An Account of the Mechanism of an Automaton, or Image Playing on the German Flute*, trans. John Theophilus Desaguliers (London: T. Parker, 1742), p. 22

235 Jessica Riskin, 'The Defecating Duck, or, the Ambiguous Origins of Artificial Life', *Critical Inquiry* 29:4 (2003), 599–633 (p. 601)

236 E. C. Spary, *Eating the Enlightenment: Food and the Sciences in Paris, 1670–1760* (Chicago: University of Chicago Press, 2012), p. 17

237 Thomas Carlyle, *The French Revolution: A History*, Vol. 2 (London: Chapman and Hall, 1837), p. 177

238 William Shakespeare, *Coriolanus* (1609), I.I 94–6

239 Ibid., I.I 146–7

240 See David G. Hale, 'Analogy of the Body Politic', in *Dictionary of the History of Ideas*, ed. Philip P. Wiener (New York: Scribner, 1973), Vol. 1, pp. 68–70

241 Bertrand Marquer, 'The "Second Brain": Dietetics and Ideology in Nineteenth-Century France', in *Gut Feeling and Digestive Health in Nineteenth-Century Literature, History and Culture*, pp. 37–54 (p. 40)

242 Robert James, *A Medicinal Dictionary*, 2 vols (London: T. Osborne, 1745)

243 James Eyre, *The Stomach and Its Difficulties* (London: John Churchill, 1852), p. vii

244 N. Henessey, 'Dietetics in Relation to Mental Culture'

245 Thomas Carlyle quoted in Hisao Ishizuka, 'Carlyle's Nervous Dyspepsia: Nervousness, Indigestion and the Experience of Modernity in Nineteenth-Century Britain', in *Neurology and Modernity: A Cultural History of Nervous Systems, 1800–1950*, eds Laura Salisbury and Andrew Shail (Basingstoke: Palgrave Macmillan, 2010), pp. 81–95 (p. 88)

246 Douglas, *Purity and Danger*, p. 50

247 Cheyne, *The English Malady*, pp. 49–50

248 Claude Lévi-Strauss, 'The Culinary Triangle', in *Food and Culture: A Reader*, eds Carole Counihan and Penny Van Esterik (London: Routledge, 1997), pp. 28–35 (p. 30)

249 Roland Barthes, 'Steak and Chips', in *Mythologies* [1957], trans. Annette Lavers (London: Vintage, 1993), pp. 62–4

250 Ben Rogers, *Beef and Liberty: Roast Beef, John Bull and the English Nation* (London: Vintage, 2003)

251 Joyce L. Huff, 'Corporeal Economies: Work and Waste in Nineteenth-Century Constructions of Alimentation', in *Cultures of the Abdomen*, pp. 39, 40

252 Frantz Fanon, *The Wretched of the Earth* [1961], trans. Constance Farrington (New York: Grove Press, 1963), p. 43

253 Ibid.

254 E. Melanie DuPuis, *Dangerous Digestion: The Politics of American Dietary Advice* (Berkeley: University of California Press, 2015), p. 147

255 Robert Harrison and Virginia Smith, 'William Banting (1796/7–1878)', *Oxford Dictionary of National Biography*

256 William Banting, *Letter on Corpulence, Addressed to the Public* [1863] (London: Harrison, 1864), p. 14

257 Ibid., pp. 7, 11

258 Ibid., p. 14

259 Galen, quoted in Christopher E. Forth, *Fat: A Cultural History of the Stuff of Life* (London: Reaktion Books, 2019), p. 54

260 Banting, *Letter on Corpulence*, pp. 11, 13

261 Kaufman, *A Short History of the American Stomach*, p. xiv

262 'diet, n.' *OED Online*, Oxford University Press, oed.com/dictionary/diet_n1

263 Geoffrey Chaucer, 'The Tale of Beryn: With a Prologue of the Merry Adventure of the Pardoner with a Tapster at Canterbury' [1400], in *Supplementary Canterbury Tales*, ed. F. J. Furnivall and W. G. Stone (London: N. Trubner & Co., 1887), p. 45; see also Jake Walsh Morrissey, '"To Al Indifferent": The Virtues of Lydgate's "Dietary"', *Medium Ævum* 84:2 (2015), 258–78

264 Michel Foucault, *Technologies of the Self*, eds Luther H. Martin, Huck Gutman and Patrick H. Hutton (London: Tavistock Publications, 1988), p. 16

265 For an overview of this research, see Ching-Hung Tseng and Chun-Ying Wu, 'The Gut Microbiome in Obesity', *Journal of the Formosan Medical Association* 118:1 (2019), 3–9

266 Elaine W. Yu, Liu Gao, Petr Stastka et al., 'Fecal Microbiota Transplantation for the Improvement of Metabolism in Obesity: The FMT-TRIM Double-Blind Placebo-Controlled Pilot Trial', *PLoS Medicine* 17:3 (2020)

267 Ian Randall, 'Personalised Diet Plan Based on Healthy Plant-Based Foods and Tailored to Your Gut Microbiome "Could Help Reduce Your Risk of Obesity, Type-2 Diabetes and Cardiovascular Disease"', *Daily Mail*, 11 January 2021; David Cox, 'Seven Ways to Boost Your Gut Health', *Guardian*, 10 September 2018; Tory Shepherd, 'Super Poo: The Emerging Science of Stool Transplants and Designer Gut Bacteria', *Guardian*, 2 January 2022; Rachel Hosie, 'Testing Your Gut Bacteria Could Be Secret to Losing Weight, Finds Study', *Independent*, 18 September 2017

268 Siobhan Fenton, 'Obesity Could Be Contagious, Scientists Say', *Independent*, 5 May 2016

269 Hilary P. Browne, Samuel C. Forster, Blessing O. Anonye et al., 'Culturing of "Unculturable" Human Microbiota Reveals Novel Taxa and Extensive Sporulation', *Nature* 533 (2016), 543–6

270 Lizzie Parry, 'Is Obesity CONTAGIOUS? Spores of Bacteria from the Guts of Fat People "Could Spread to Healthy Individuals"', *Daily Mail*, 4 May 2016

271 Jana Evans Braziel and Kathleen LeBesco (eds), *Bodies Out of Bounds: Fatness and Transgression* (Berkeley: University of California Press, 2001), p. 2

272 See Charlotte Cooper, *Fat Activism: A Radical Social Movement* (Bristol: Intellect Books, 2016)

273 Ken Albala, 'Weight Loss in the Age of Reason', *Cultures of the Abdomen*, pp. 169–83 (p. 170)

274 D'A. Power and Kaye Bagshaw, 'William Wadd (1776–1829)', *Oxford Dictionary of National Biography*

275 William Wadd, *Cursory Remarks on Corpulence; or, Obesity Considered as a Disease: With a Critical Examination of Ancient and Modern Opinions, Relative to Its Causes and Cure* [1810], 3rd edn. (London: J. Callow, 1816), pp. 21, 24–25

276 William Hunter (1718–1783), Scottish anatomist and physician, known for his pioneering work in obstetrics

277 Marcello Malpighi (1628–1694), Italian biologist and physician famed as the 'founder of microscopical anatomy'

278 Herman Boerhaave (1668–1738), Dutch botanist, chemist and physician

279 Gerard van Swieten (1700–1772), Dutch physician and doctor to Holy Roman Empress Maria Theresa

280 Wadd, *Cursory Remarks on Corpulence*, pp. 10–13, 59

281 Ibid., p. 14

282 Ibid.

283 Michael Stolberg, '"Abhorreas pinguedinem"': Fat and Obesity in Early Modern Medicine (c. 1500–1750)', *Studies in History and Philosophy of Biological and Biomedical Sciences* 43:2 (2012), 370–8 (p. 375)

284 Samuel Smiles, *Self-Help; with Illustrations of Character and Conduct* (London: John Murray, 1859), p. 18

285 Sander L. Gilman, *Fat: A Cultural History of Obesity* (Malden: Polity Press, 2008)

286 See S. Rosenbaum, '100 Years of Heights and Weights', *Journal of the Royal Statistical Society* 151:2 (1988), 276–309

287 Francis Galton, *Hereditary Genius: An Inquiry into Its Laws and Consequences* (New York: D. Appleton and Co., 1870), pp. 1–4

288 Kevin Donnelly, *Adolphe Quetelet, Social Physics and the Average Men of Science, 1796–1874* (Pittsburgh: University of Pittsburgh Press, 2015)

289 Adolphe Quetelet, *A Treatise on Man and the Development of His Faculties* [1835] (Edinburgh: W. & R. Chambers, 1842), p. 14

290 Your Fat Friend, 'The Bizarre and Racist History of the BMI', Medium, 15 October 2019

291 Ken Albala, *Eating Right in the Renaissance*, p. 139

292 Kathleen LeBesco, *Revolting Bodies?: The Struggle to Redefine Fat Identity* (Amherst: University of Massachusetts Press, 2003)

293 Tessa E. S. Charlesworth and Mahzarin R. Banaji, 'Patterns of Implicit and Explicit Attitudes: IV. Change and Stability from 2007 to 2020', *Psychological Science* 33:9 (2022), 1347–71

294 'Bile Beans', *Brisbane Worker*, 6 June 1899

295 Ibid.

296 'Bile Beans', *British Medical Journal* 2:1653 (1903)

297 William Arbuthnot Lane, *The Operative Treatment of Chronic Intestinal Stasis* (London: James Nisbet, 1915), p. 55

298 Ellen G. White, *Education* (Oakland: Pacific Press Publishing Company, 1903)

299 Ana Carden-Coyne, *Reconstructing the Body: Classicism, Modernism, and the First World War* (Oxford: Oxford University Press, 2009), p. 158

300 Ina Zweiniger-Bargielowska, 'The Culture of the Abdomen: Obesity and Reducing in Britain, circa 1900–1939', *Journal of British Studies* 44:2 (2005), 239–73 (p. 239)

301 Eustace Hamilton Miles, *Self-Health as a Habit* (London: J. M. Dent & Sons, 1919), p. 25

302 Nathaniel Newnham-Davis, *The Gourmet's Guide to London* (New York: Brentano's, 1914), p. 77

303 Ana Carden-Coyne, 'American Guts and Military Manhood', in *Cultures of the Abdomen*, pp. 71–85 (p. 72)

304 Philip Rucker, Josh Dawsey and Damian Paletta, 'Trump Slams Fed Chair, Questions Climate Change and Threatens to Cancel Putin Meeting in Wide-Ranging Interview with The Post', *Washington Post*, 27 November 2018

305 Robin M. LeBlanc, *The Art of the Gut: Manhood, Power, and Ethics in Japanese Politics* (Berkeley: University of California Press, 2010)

306 'Commons Outrage', *The Globe*, 25 June 1909; 'Suffrage Protest', *Morning Post*, 25 June 1909

307 Marion Wallace Dunlop, 'Address to the WSPU', *Votes for Women*, 1 August 1909

308 'Release of Women Suffragists', *The Times*, 14 March 1908

309 Edith Ward, 'Review of *Animal Rights* by Henry Salt', *Shafts*, 19 November 1892

310 Graham Greene, 'Alas, Poor Maling' [1940], in *Twenty-One Stories* (London: Vintage, 2001), pp. 69–72

311 Ali Khanbhai and Daljit Singh Sura, 'Irritable Bowel Syndrome for Primary Care Physicians', *British Journal of Medical Practitioners* 6:1 (2013)

312 crowdresearch.uwa.edu.au/project/noisy-guts-project/

Index